PENGUIN SPECIAL
**An African Winter**

Preston King fled the United States in 1961 following his arrest and conviction by federal authorities for early involvement in the Southern Civil Rights Movement. He has been exiled from America for twenty-five years.

He has taught philosophy and politics since he was twenty-two, first at the London School of Economics, then at Keele and Sheffield Universities. Since 1963 he has worked in and regularly visited Africa, lecturing for four and six years respectively in Ghana and Nairobi and serving in Cameroon and elsewhere for shorter periods. During his time in Nairobi, he derived most pleasure from setting up the Diplomacy Training Programme at the University.

Preston King has wide interests, as reflected in his books, which include *Fear of Power* (1967), *The Ideology of Order* (1974), *Toleration* (1976), *Federalism and Federation* (1982) and *The History of Ideas: An Introduction to Method* (1983). For some years he held the chair in political science at the University of New South Wales in Sydney and he now holds the chair of politics at the University of Lancaster. He is currently working on a major study of ethnicity and politics.

*Preston King*

# AN AFRICAN WINTER

*With a Note on Ecology and Famine*
*by Richard Leakey*

Penguin Books

Penguin Books Ltd, Harmondsworth, Middlesex, England
Viking Penguin Inc., 40 West 23rd Street, New York, New York 10010, U.S.A.
Penguin Books Australia Ltd, Ringwood, Victoria, Australia
Penguin Books Canada Limited, 2801 John Street, Markham, Ontario, Canada L3R 1B4
Penguin Books (N.Z.) Ltd, 182–190 Wairau Road, Auckland 10, New Zealand

First published 1986

Made and printed in Great Britain by
Richard Clay (The Chaucer Press) Ltd, Bungay, Suffolk
Filmset in Monophoto Plantin by
Northumberland Press Ltd, Gateshead, Tyne and Wear

In memory of my late friend
Okot p'Bitek
who wrote, with knowledge, of pain

And in praise of the work of
OXFAM
who have done so much, so efficiently,
to diminish its hold

# Contents

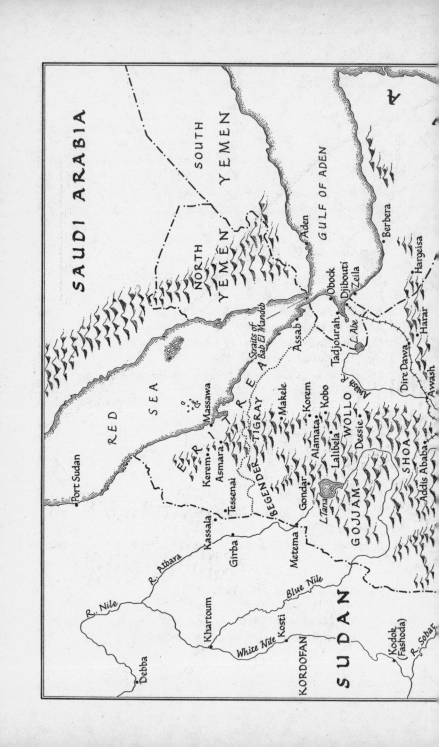

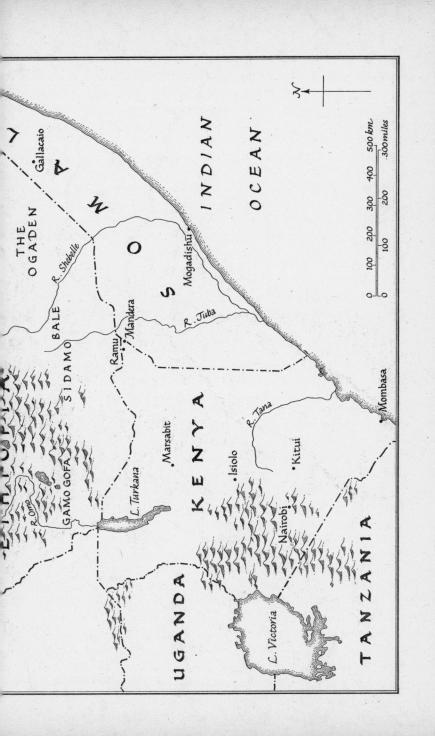

# Preface

For Africa, independence was a boon, met with joy. But it imposed a burden, increasingly sustained with sorrow. Each new, fragile state, based on an uncertain and vulnerable tax base, was insecurity itself. With severely limited resources and prospects, they were all swiftly exposed to regional and international competition. They faced three great difficulties.

The first, though not necessarily the most important, had to do with the structure of economic relations between former colonial centres and colonial peripheries. This unbalanced and indeed exploitative structure persists in the present high levels of African debt to the northern states. The second difficulty had to do with natural conditions, such as friable tropical soils, uncertain rainfall and pest infestation. The third problem related to the quality of home and foreign policy, especially where these inclined towards conflict, as they do increasingly. The crisis in which Africa is currently enmeshed is a combination of these and associated problems.

African states are independent, not autonomous. They form an inescapable part of the global economic and political system. What African leaders do in Africa affects African peoples, but what American leaders do in Washington also affects African peoples. The consequences of decisions taken in the City of London, on Wall Street or in the other financial capitals of the North are immense, and yet not obvious to lay observers. It is important to avoid the trap of Europe-bashing, but it is misleading to lay blame 'fairly and squarely' (which means mostly) at the feet of African elites. It is important to avoid falling overboard in the promotion of no-holds-barred, free-enterprise solutions, but there is no point in hot-ballooning it into the heaven of simplistic and all-encompassing statist solutions. No problem is to be explained exclusively by reference to the victim: we all have our own share of responsibility to bear.

The one, central change in attitude I would wish to see on the African side is a weakening of the tendency to avoid (for whatever reasons) sharp debate of public policy. The one, central reversal I would wish to see take place on the North Atlantic side is of the tendency ostentatiously to ignore the extent of northern involvement in the continuing African *débâcle*. The level of palpable *official* indifference is appalling. Africa's is a global problem and requires to be met on a global basis, leaving ideological cant and other forms of misplaced self-righteousness to one side.

One of the most important tasks of the great powers in the African continent is to pursue the restoration of peace with justice by every means. They must be dissuaded from deliberately fomenting or acqui-escing in local conflicts in order to secure strategic or other advantage. Africa needs less, not more, conflict, and it is only a pulling-back from the fearful misuse of limited resources that will allow constructive responses to critical problems; as we survey the continent, however, what we observe is not more peace, but more war.

To the north, Morocco and Polisario do battle for Western Sahara. Libya imposes herself upon Chad. Burkina Faso and Mali are swept up in an absurd engagement along the whole of their common frontier. Egypt enjoys an uncertain peace, contending with a destabilizing Israeli irredentism to the east and with a threatening Libyan adventurism to the west. In the Sahelian arc that runs from the Atlantic to the Red Sea, not a single state (Mauritania, Mali, Burkina Faso, Niger, Chad, Sudan, Ethiopia) has been spared a *coup d'état*. Civil war has crippled most of them. And civil war has settled on Uganda – wedged between Sudan and Tanzania, between Zaire and Kenya – the cumulative death toll over the past decade passing the 600,000 mark. To the south, the apartheid Republic stumbles from crisis to crisis, dispatching its troops to the outer limits of its regional empire, to smash into Angola, Botswana, Mozambique and other states. It complains querulously about the lack of democracy abroad, but it is stricken with paralytic fear that democracy may overtake it at home.

The images of dying children remind us of the tie between these deaths and drought. But with every increment of peace, just that much more of vital energy and resources can be allocated to contain drought and associated problems. Famine is usually at its worst in states that are at war.

\*

This book was conceived in January 1985 and completed a year later. A few people contrived to make life difficult. Most were fantastic in their helpfulness. Those friends, associates and colleagues who have helped will know quite well who they are. The vast number of people involved, together with their institutional affiliations, have inspired me to omit, for good and ill, a naming of the names of the virtuous. What is perfectly clear is that, without the help of so many people, such a book could not have been completed – or even indeed begun.

This book is not written as an academic essay. The material and experiences from which it is drawn are so complex that to supply the normal range of citations seemed impracticable. The final chapter (14) is more open-ended and argumentative, and it seemed necessary to attempt to supply references there in a somewhat more systematic way.

# Part One  DROUGHT

# On Ecology and Famine

by Richard Leakey,
Director, National Museums of Kenya

Africa is now generally recognized as being the birth-place of our species – if you like, the veritable 'Garden of Eden'. Our earliest ancestors evolved in Africa and became human. They learned to use their expanded brains to fashion technology. By so doing, they established a pattern of life that was to be successful for close to two million years. These people were able to move throughout Africa and beyond. One million years ago, human ancestors were beginning to move across what we know today as Europe, Asia and the Far East. Much, much later, perhaps only in the past seven thousand years, technology as applied to food production resulted in a dramatic increase in population numbers. Nation states and civilizations sprang up across the world. With the development of agriculture, famine and mass starvation confronted our species as a new factor. Small mobile populations would have known hunger, but famine is unique to large populations with restricted dietary options. Put more simply, famine is a consequence of an agricultural way of life that has failed because of drought.

We now know that in Africa, a million and a half years ago, our ancestors were large and well built. This stature reflects a good diet, and I have no doubt that early populations of people lived in the very best places, where wildlife and foods were plentiful the year round. It is in these areas, much later, that people turned to agriculture, pushing nomads and hunter–gatherer communities to the less bountiful areas. While crops do well, people thrive and multiply; but if crops fail, there are too many people to move and there are insufficient resources to enable a return to living off the now impoverished land.

An examination of the past can be instructive, and there are some important lessons to be considered from the current crisis in central Africa, Sudan and Ethiopia. Over the past decade there has been a significant decrease in rainfall, with attendant changes in the environment. One of the most dramatic illustrations of this can be seen in the present water levels of the African lakes that have their catchment areas in these regions. Lake Chad in central Africa is at least seven metres lower than it was fifteen years ago, while Lake Turkana (the former Lake Rudolf) in northern Kenya is six metres lower. Lake Turkana receives virtually all of its water from the Omo River, which drains south from Ethiopia's highlands. Geological records indicate that these lakes have in the past completely dried up save for a few swamps, while at other times they have been as much as fifty metres higher than they are now. The current drought in Ethiopia is represented today by a mere six-metre drop in water level at Turkana. If the lake drops another ten metres over the next ten years (as it well might) we can expect a disaster which would make the present famine pale into insignificance. In historic times, the lake has been at least ten metres lower than today, and old people well remember islands that are yet to reappear but upon which they grazed livestock and took refuge during raids at the turn of this century.

We don't hear of great famines on the scale of 1985. This may be because there was no television and other media coverage or because there was no widespread famine despite the occurrence of drought. It is likely that, when there was drought, people moved to new pastures. International frontiers did not exist and large language-related groups were able to exploit the available resources, drawing upon ages of experience. Today, people cannot move except as refugees. They have to cross frontiers and move from one political system to another. Populations are larger, herds of stock are bigger and the old wisdom is failing. Modern emergent nation states have accepted the most up-to-date technology for 'death control' but have failed to take on appropriate systems of 'birth control'. We prolong life with elaborate and expensive programmes of medical care, but scant attention is given to preventing birth through family planning. Leaders speak knowingly of the carrying capacity of the land in terms of domestic stock and crops, yet they fail to see the same equations when looking at their own people and the capacity of their own country.

It is not possible to blame the former colonial countries for the drought; equally, it is not realistic to view the drought as abnormal. Africa is as familiar with drought as the northern European and American states are with winter. To sustain life through winter months, people plan for their needs and adapt their life-styles. Surplus food and stock are stored, and herd sizes are kept to numbers that can be sustained. We in Africa must do the same, because drought will continue.

The traditional way of life used to work; populations and herds were mobile and could be moved to empty lands as the need arose. Given that this is now impossible, and the traditional way of life has gone, African governments must be responsive to the realities of the present.

African leaders – politicians, generals, scientists and business people – must recognize that, while the problems arising from drought can be treated, in the long run prevention is better (and cheaper) than cure. Inviolate political frontiers created by the colonial powers and sustained by economic and ideological theory are as great a problem to Africa's well-being as are the cycles of climate which wreak havoc with carefully considered plans. The climate cannot be modified, but our attitudes to human affairs and development can be. We must move forward to an Africa where everyone can develop a productive life and where the sum of everyone's efforts will strengthen the continent's ability to contend with natural long-term cycles in rainfall.

There has to be a balance between people's basic needs and the productivity of their 'system'. If numbers of people increase, productivity must be increased at the same rate. In traditional Africa, there probably was an equilibrium: people thrived and population density was related directly to the carrying capacity of the land. In modern times, medical intervention both prolongs and saves lives; veterinary care and piped water increase the numbers of domestic animals; social values lead to a greater frequency of childbirth. The opportunities and facilities for this greater biomass have changed little, however, from traditional times, and the same 'way' of life often persists. Ecological balance, like any equation, has to be a balance for success. We have increased the requirements but not the resources, and starvation results.

In my view there is much that can be done. I believe that it is imperative to recognize that the traditional way of life cannot work if sustained development is the objective. If people are to have the benefit of modern technology, they must accept the discipline of a way of life that can sustain these new advantages.

The massive emergency aid that has been pouring into Africa to feed the starving is only a short-term palliative, not a cure. We must recognize that the real causes of famine go much deeper and that we, too, must go deeper. It is true that we require development aid on a vast scale, but it will be to no avail without the courage and determination to use it properly.

# 1 Addis Ababa: Viewing from a Distance

*January 1985.* As the Ethiopian jet lanced the cloud, Addis Ababa burst suddenly into view. Effervescence is not a part of the city's nature. The land was green, but a faded green. The eucalyptus, imported from Australia in the previous century, displayed a certain robustness, but they were few in number. 'What's growing here?' I asked the air hostess, as we waited for transport into the city. The smile, worn so convincingly at six miles high, could not be found. 'They're all too tired,' she said rather indirectly. 'Nothing's growing, really.'

I finally secured a taxi for 10 birr, about $5. In Revolution Square, an enormous portrait of Ethiopia's President faced an equally enormous poster of Marx/Engels/Lenin, aligned in profile. We tried the Ghion Hotel, with its low rates and authentic atmosphere, but there was no room. We continued towards the Hilton, past an immense statue of Lenin on Menelik Avenue, opposite the Economic Commission for Africa (ECA) – not there on my previous visit. Immense spaces seemed to separate ceremonial structures: portraits, obelisks, slogans, palaces, broad boulevards. Hovels everywhere, as quiet as you please.

Having booked in, I took another taxi along Development Through Cooperation Avenue (the name had changed) to the famous lion cages. Two lads hailed me along the way and told me their names were Tewodros and Abil. They wished to speak English and followed me round the cages. A sort of picket fence, painted rust colour, had been erected at a distance of perhaps six feet from the animals' perimeter. There used to be three to four dozen animals kept here; now there were only fourteen adults and five cubs. The picket fence seemed oddly out of place, but I remembered how, on first visiting the lions, more than ten years before, my companion had almost been killed. In the twinkling of an eye, a lioness, for some indecipherable reason, had sprung from the rear of the cage, her paw missing my stationary friend, back turned, by inches. And the new fence? 'No,' the men said, 'no deaths or injuries. Just dangerous.' Somehow the lions looked less dangerous now than then.

I relinquished my taxi and began my walk back south, first to the National Museum, on King George V Street: some names do not change. Two young soldiers stood outside, seemingly indifferent to all that passed. Tewodros and Abil were still with me, then fell behind. When I looked round, they had their arms casually extended above their heads, inviting the soldiers, who turned out to be guards, to search them. They were frisked, equally casually, and admitted into the museum. It used to publish a great deal regularly, but now had nothing later than 1982.

We left and made our way to the Economic Commission. It was like running a gauntlet. Tiny children were everywhere. Abil said they were going to school, where there were three shifts: the first from seven o'clock in the morning, the second from eleven, the third from three in the afternoon. Tewodros volunteered that there was a fourth at night, for adults. Tewodros and Abil attended the first. We continued, enveloped in a quiet hum. No shouting. London streets, not to speak of Lagos or Freetown, are far noisier. A mass of folk passed virtually empty stores and rare fruit stands with oranges and oblong water-melons. There were book shops with few books, all old: Marx, Engels and Lenin, in paperback, with Moscow imprints.

The two boys had made it clear that their object was not money, but English. And they liked Americans, who were helping Ethiopians. It was so bad here. There was nothing to do. I left them with a contribution to buy writing paper, which was very expensive. They smiled handsomely.

In the E C A I was accosted by a blind man led by one who was sighted, by a seated woman with outstretched arm over whose legs I had to step, by a well-dressed young man who prided himself on his cleverness in divining that I was a black American and by a child who charmingly insisted that he should be allowed to shine my suede, open-toed sandals. Having made my way into the building, I appropriated the chief documents clerk, Mr Kessela, who quickly and helpfully located what was required. He sucked in his breath in characteristic Amharic fashion, assenting to one request after another for documents in his custody.

The Organization of African Unity was remarkably different. The Director, Peter Onu, would not be back until the following Monday; he was on tour in the Sahel. Most of those whom I had hoped to see were not available, and the organization closed for lunch between 12 and 2.30. I was able to see some of the documents I needed, but the wheels of the organization seemed to turn very slowly. The map of Africa at the entrance to Building A envelops the wall, a map fashioned from copper or brass, perhaps fourteen to eighteen feet high. Each state is physically set off from every other, but no names are inscribed on them: the viewer must work this out for himself. I halted for a moment, rehearsing the names, starting from the south, with South Africa (Azania), Lesotho, Swaziland, then Botswana ... but Botswana appeared to be enveloped by Zimbabwe, or Zimbabwe by it. I asked one of the ladies working there if I was hallucinating, or had Botswana and Zimbabwe been merged? A passing male colleague was called in, and the idea that there may have been a mistake was rejected. Then the excuse was offered that it was an *old* map, executed in 1962. 'Has it actually been like that ever since?' 'Well, yes.' We concluded by agreeing that something should be done about it, but in such a way as to suggest that there were other matters of greater moment to attend to.

At the Ethiopia Hotel I saw the World Vision relief workers who had established themselves on the first floor. I spoke with a pleasant young American, who pointed me to one of his colleagues. I made for the appropriate office. The occupant, I was advised, was 'away', so it seemed as well to have a bite before returning to the ECA for an afternoon appointment. I went down to the dining room, as if into a sunken garden; but rather than exotic plants, the eye fell upon plates and steel tureens. Transfixed, I approached the array of connected tables; they formed a horseshoe, holding up to view diced fish, plates of beef, mutton, *njera* and *wat* (Ethiopia's 'bread' and stew), potato salad, beetroot, coleslaw, chicken, peas, beans, bread, fruit and cakes. Around it slowly gyrated a queue of smiling relief workers, foreign and Ethiopian. I invited myself to a table where sat three others; on my left was a large woman, another American. Next to the buffet an Ethiopian sat playing the piano in the never-ending, but endearing, tinkly manner of the country. For the large lady his playing did not enhance the meal. 'All in the same key,' she remarked above the clatter of metal on plate. 'Sometimes he does three to the bar, sometimes four. But they play by ear. That's what happens when you play by ear.' She had been in Ethiopia for a brief period, was a secretary with one of the relief organizations and could not wait to get home. The bureaucracy, she reckoned, was 'worse than ours'. Yet another Ethiopian student wanted nothing so much as to improve his English: there was an orphanage which he would be pleased to show me ... if I had the time. What was I doing at 6 o'clock?

Ethiopia is famous for its silverware, especially its crosses. They seem to be in every shop window in Nairobi, but are no longer so obvious in Addis Ababa. 'What has happened to all the silver?' I asked an acquaintance who was driving me about.

'We owe the Russians everything,' he said. 'So they take everything. We pay them with our silver, with our gold. Our coffee, too. It all goes there.'

Ethiopia would have disintegrated after 1977 had it not been for a massive airlift of Soviet support to stem the guerrilla tide. The country is now heavily in debt. The morning's paper mentioned none of the many private Western relief agencies at work here, apart from an editorial which cited the International Red Cross. On the front page, top right, was a photograph of Comrade Andrzej Konopacki delivering '165 metric tonnes of nutritious food from the government and people of Poland'. At bottom left was a photograph of Comrade Ambassador Georgi Kassof of Bulgaria surrendering more 'wheat flour, nutritious food' to Ethiopia's Comrade Berhanu Deressa. One writer canvassed the idea of massive aid to Africa along the lines of the Marshall Plan, partly on the grounds that 'many European nations owe a lot to the African people who have in one way or another contributed to the development and progress of that continent'.

On the following day I took a taxi to the market, the Mercato. The taxi driver, anxious for trade, quietly warned me that this was an unsafe place in these unsettled times and that he should wait for me. Would I be long? I begged him not to worry. He drove away, shaking his head slowly, looking for clients.

Everywhere in the Mercato, food and cloth abounded. Sacks of grain seemed to have been stacked in every conceivable corner. All of these things were costly. There were skinny people everywhere, and the lame, and the blind. On the surface, misleadingly, as in Bombay, merchants seemed comfortable, their supplies heaped up in piles. In Bombay, plump mothers with newly born children beg at cars that pass along the seafront. In Addis Ababa it is more stark. The blind, led by the sighted, both thin, beg at the traffic lights. In Bombay, happily, one can at least sleep out under the stars, even at the risk of having one's toes nibbled by scurrying rodents on the take; to sleep out in Addis is to invite death by cold.

The statistics for Addis Ababa reveal a high incidence of tuberculosis, pneumonia and other respiratory infections in the population of 1·5 million. In the country as a whole, life expectancy at birth in 1984–5 was in the mid-thirties; the figure is marginally higher for Addis (and for Ethiopia's few other cities). In Addis, diseases of the gut, parasitic infestations, account for about two-fifths of all deaths, with a preponderance of child victims. There is a great deal of malnutrition, but that, unless very severe, is not so easily seen. There are serious problems in the city, as there would be anywhere where one dares to aim at reducing the infant mortality rate from in excess of 150 per 1,000 to 50 per 1,000 by the year 1990. This simmering misery is a matter for study; it is held back, it does not exactly erupt into view. Addis Ababa may well be the eye of the storm; there is famine in the country, but the city is not the place to find it. For perceptible tragedy of a kind that any tourist can understand, it is necessary to explore the hinterland.

## 2  Korem: Panning In on Death

Two hundred and twenty-five miles north of Addis Ababa, in the Ethiopian province of Wollo. A child without a name stares in quiet horror. Its eyes are huge, their muddy whiteness exposed by the contraction of surrounding flesh. The skin of the face is taut, the mouth too, both retreating from a substratum of teeth and skull. The flesh withers almost visibly, like water at low tide rippling across stolid rocks. The impulsive exuberance that belongs to infancy is absent. The motion that this child achieves is genuinely disembowelling, but, in keeping with the setting, there is not a trace of violence about it. This slow, soundless seepage from the lower colon is apportioned in tediously minute doses and signals the even deceleration of life towards early, implosive collapse.

A drip performs as intended, taped to the baby's forehead. By this means, the nurse's aid force-feeds the child through its nose. The child's torpor seems bizarre. Legs and arms and chest have quite wasted away. The bony head and hands and knees appear huge beneath the taut skin that twists the limbs at defensive angles. The eyes provide a singular contrast to this miniature bag of bones: they bulge above it and appear to fight against it; they give the lie to sleep. The long curling lashes blink wanly, half in surrender, half perhaps to keep the evil moment at bay. But the drip cannot revivify the mummy. It cannot smooth the wrinkles from the skin. As little as it brings, it is too late, and indeed it brings far too much for a terminated life to feed on. Nothing more can be done. Instinctively, one looks away.

To remove one's eyes from the dead is not to take them from death, which is everywhere. There are tears, as one might expect, but not many: there is so little water to spare. There are children everywhere: 7,500 is the figure given, about three in five of whom are being intensively fed, six times a day. Those so nurtured are severely underweight, and every effort is made to bring them back to par. When they reach what may be called 'normalcy', they are of course discharged, to re-enter upon a degenerative process which will either return them to this shelter or (more likely) conclude their stay on earth. To survey the living child is not to look out upon life, but always upon diminution and loss of life. These children do not play; few have sufficient hope or strength to muster a smile. Their small bodies have limited resources. While children require more food than adults – three times the calorific intake – malnutrition and diarrhoea will cause them to absorb less. Sometimes, it is only with the greatest difficulty that they can be persuaded to eat at all – hence the drips. They eat little, slowly, laboriously and without relish. These children are highly vulnerable; they die in great numbers; they are thus accorded priority treatment. At this altitude, and with little cover, there is much flu, bronchitis and pneumonia. There is whooping cough and tetanus; diarrhoea constitutes a way of life. Those who ingest too much invite severe stomach cramp and an onset of vomiting. Those who ingest too little (which is why they are here) simply continue to starve.

Survival, if achieved, is not life. If it is life, it is at zero point. Survival is what is left after all the joy is taken out. The wasting away that the children endure the doctors call marasmus. We call it starving. To survive starvation will cost something, and usually a great deal – such as weight and height and possibly even sight. Survival may be purchased at the cost of sound limbs and vigorous minds. Stunting, wasting, blindness, lameness, diminished intellect – all prefigure an early death, and each represents a sort of mortgage with which a child may be burdened in securing an extension of life.

This mother who sits on the ground, her back resting against a pole, is almost as lifeless as her child. There is a pungent odour. She is not clean. The lice outside, the worms inside, are feeding well, though quietly, out of earshot. They draw blood and she cannot respond. She stares, which requires no effort, since she stares at nothing, unless at a distant and fatal horizon, which almost everyone here seems adept at discerning. Her bones have achieved an ascendancy over her flesh. The breasts hang flat, outlined against the thin rag that covers her chest. The mouth is ulcerous, slightly agape, a fly at the corner. There is a pulse, but it is slow; body heat, but low; the blood is anaemic. She has often been pregnant. She has planted and sowed and walked and worked for her family, and is none the less so diminished now that she, herself, is all that is left. The nurse's aid says it took a three-day walk for her to reach Korem. Others have taken longer. This mother sits. She is only a child herself: ancient, but not much over eighteen. Imperturbable, she waits her turn. There are no riots here, no explosions, only despair and easy resignation.

And yet so many Ethiopians have made their way to Korem, from all points of the compass. Extraordinary determination is reflected in the accomplishment. The point to which they have come is high on the plateau, above 8,000 feet, freezing at night, enveloped by mist at dawn, swept by strong, often biting, winds. Korem is a vast, Faustian arena, rolling away from the eye, brown, parched, ballooning into dust. The whole region is broken, craggy, inaccessible: only 2 per cent of Wollo can be reached by car, but these people have come, both in hope and desperation. Here they expect to link up with the lifeline of a collapsed civilization – the major arterial road which leads north from Addis Ababa – through the provinces of Shoa, Wollo, Tigray and Eritrea – to debouch finally at the port of Massawa on the Red Sea. To either side of this road, as it passes through Wollo, lie pools of hope: the towns of Dessie, Kombolcha, Lalibela, Kobo, Alamata, Korem. The planes which bring relief workers and visitors cannot land at Korem. They make their way to Alamata in the valley, fifteen miles below. It is from there that passengers and supplies may be trucked up the winding mountain road that leads one in, past the soldiers who stand sentry all round the perimeter.

The 'normal' population at Korem was said to be 7,000, but by March 1983 the figure had risen to 17,000. After March, numbers jumped to 38,000. Within two months (by May 1983) the refugee inflow was given as 45,000. By July 1984, according to the United Nations, there were 102,000 people. The camp at Korem had limited facilities, food and personnel. There were many more people requiring care than could be given it. There were more people being turned away than were, or could be, helped. For this reason, largely, the figures are unreliable. When Perez de Cuellar, UN Secretary-General, visited Korem in early November 1984, the number of displaced persons at the feeding centre was reported by the press as 35,000. This was the figure given for the previous May, when it was also reported (by the same source) that an additional 110,000 people were registered with the government agency (the RRC) in Korem for emergency relief. The number of displaced persons at the feeding centre in December 1984 was variously given as 50,000, 45,000 and 57,000 (the last two figures by the *Washington Post* on 6 and 15 December respectively). *The New York Times* (9 December) gave 40,000. In January of 1985 there emerged some general agreement among press personnel that the numbers amounted to about 60,000. But the number remains uncertain, and in June 1985 it was given as 50,000.

The overall picture of Korem is one of swelling population, certainly from 1982, but with various irregularities depending upon season and security. There were insufficient stores, shelter and personnel, so that only a proportion of those in need could actually be helped. There was a great and rising crush of people, coming and going, in search of aid, despairing of finding it, moving out of Korem to other towns, or as far as their physical resources would allow. At first people died because there was simply not enough food. In November 1984, the food supply haltingly began to improve, but death from cold, exposure, disease and exhaustion increased. Interwoven with these developments, security was tenuous, despite the presence of soldiers (sporting AK-47 rifles) and armoured vehicles. Supplies were interrupted. The road leading north from Addis Ababa was not safe. Many of the towns around Korem (and indeed Korem itself) had been attacked. Alamata and Lalibela had been overrun. Ethiopian troops drove many refugees south from Korem to resettle them. At least one convoy of seven trucks was intercepted by Tigrean rebels, the refugees off-loaded and the vehicles destroyed. The conclusion is not that there were no more than 50,000 or 60,000 at Korem, but that no attempt could be made to save more than this.

People in Korem were dying in swarms. Had they not come, they would have died. Having come, they died all the same. Half of the 60,000 said to be here at the beginning of 1985 had no cover. There were makeshift strategies, of course, most notably the shallow rectangular pits where starving, freezing people might shelter, plastic sheeting pulled overhead, those so protected sitting back to back, or face to face, or head on lap, or body on body, all with a view to conserving body heat, against the spluttering, coughing, sneezing and wheezing that enveloped them. There were blankets to be distributed, but not sufficient; and doctors in attendance, but only a few; increasingly at least there was food, if not enough. Resistance was low. Anything could push a body over the edge at night, and you would find them there in the light of morning.

Death descended upon this place like a flock of magpies, now plucking this one, now that. The precise number of dead cannot be told. It may be that the emotion of the individual death distracts from a coldly systematic body-count. Again, we must take the picture in the round. What is clear is that, as the numbers of refugees and victims of famine increase, as the inflow to Korem swells, and the food or the blankets or both fail to keep pace, so the death count moves inexorably higher.

In December 1982, there were only eight deaths at Korem. In January of 1983, the figure moved to twenty-three. In February, to fifty-two. In March, to fifty-eight. The pace quickens. In the first week of April, the death toll nearly quadrupled, with forty-seven. Henceforth there is greater confusion, and guerrilla attacks also take their toll.

In April 1984, three hundred children died from an outbreak of measles. French medical volunteers were brought in. On 2 October 1984, the Parisian daily *Le Monde* suggested a maximum death toll at Korem of 'fifteen to twenty persons' per day, or 600 per month. The same newspaper, three months later (15 January 1985), reported a figure (again for October) of 1,800, fifty-eight deaths per day.

On 23 October 1984 Michael Buerk of the BBC reported a death rate of one 'every twenty minutes' – seventy-two per day, or 2,232 per month. Mohammed Amin, who had accompanied Buerk as cameraman, wrote in the Nairobi *Daily Nation* of 25 October 1984 that he had observed in Korem ninety to one hundred deaths daily, which would yield a maximum figure for October of 3,100 deaths. The estimate of the United Nations Children's Fund (UNICEF) for that month was the same.

Andrew Hill gave a figure of between fifty and one hundred deaths per day at Korem in November, while *Le Monde* gave a lower figure of thirty per day. One visitor, Graham Hancock, surmised (early December 1984) that there were 'perhaps 300 corpses' a day. Another visitor, Clifford May of *The New York Times*, put the figure for this period at thirty a day. The death toll in late January 1985 was given as well over fifty per day or 1,550 for the month.

The feeding centre at Korem was a beacon of life to the Ethiopians who drew near, but it also beckoned to death. Conditions were crowded and insanitary. In an unpublished report from Atlanta's Centre for Disease Control, it was claimed that cholera had been present in the region 'since at least early January 1985'. A spokesman for the Ethiopian government denied the existence of cholera in the country. In April 1985, Dr Leguillier, leader of the French medical team at Korem (the group Médecins sans Frontières, or MSF), had said that it did exist. He was quoted by reporters to this effect, while other relief workers expressed the fear that to speak out might lead to their being kicked out.

In March 1985, in a camp of comparable size in neighbouring Somalia, it was confirmed that an outbreak of cholera had in the space of three weeks killed 1,600 people. Cholera was also suspected across the border in Djibouti, and there were claims that it had struck at several camps besides Korem, especially at Makele in Tigray Province and at Ibnat in Gondar Province, the total loss of life being roughly estimated at 1,000. On 17 May, Dr Leguillier claimed that the outbreak at Korem had been contained. People did not stop dying for all that. By July, the MSF was able to declare, matter-of-factly and without further contradiction, 'We now have an average of sixteen cases of cholera a day.' The scale of death at Korem was unprecedented. The slate was being wiped clean, mostly of the very young and of the old; there were not many here who were *very* old to start with.

There are many ways of dying. Some retch and defecate as they go. Some fight bitter, protracted battles against the pangs of hunger; remorselessly they chase here and there for scraps, for offal; they are prodigal with such energy as they have in the effort to renew it; they burn off their fat, and then the flesh itself; they revolt to the end against the very idea of an end; but then, suddenly, the heart goes. There are some who could fight that have no fight; who appear affronted, appalled that human life should reach such a nadir, and shut their eyes to what is going on about them, to the rags, the stench, the humiliation, the animality. They lie on the ground in a ball, as though indolent and asleep, so that when they die the doctor who looks closely may find it impossible to identify a specific, physical cause of death.

It must also be noted that there was new life at Korem. In counterpoint to death, even in this infernal setting, yet more infants were born. Granted, these infants were few in proportion to the numbers of persons of childbearing age. Granted, too, they were as diminished and malnourished as the anaemic and prostrate child-mothers who bore them – with even less hope of survival and virtually none of a future with dignity. But there they were, adding their puny weight to the problem. It is pardonable for the distressed observer to think famine to be nature's way of foreclosing on runaway reproduction: pardonable, perhaps, but mistaken. For against such a calamity as famine, with the grotesque insecurity it breeds, the struggle for life is engaged with redoubled energy. Famine, far from checking population growth, seems to put a premium on it. Animals whose young die in large numbers breed in large numbers. Birth-control is for the rational, the secure.

# 3 Ethiopian Panorama: Misery at Large

It is obvious that, in order to live, people require food, shelter and clothing. It is equally obvious that some have more of these things than others. 'Rich' individuals may drown, be struck down by lightning, become lost in forests – but they do not die of drought or famine. Their purchasing power lifts them over such hurdles. More than this, they often positively benefit from the fact that others fail: their advantage allows them to buy more for less. The destitution of some usually has the effect of enriching others.

Countries are in important respects like individuals. Those that are 'rich' have greater reserves or surpluses than those that are not; where disaster strikes, the rich are more resilient, better able to spring back. Ethiopia has little such resiliency. It is a decidedly poor country, one of the poorest in the world. Its gross national product (GNP) for 1982 was US $140 per capita, by contrast with the United States' GNP per capita of $13,160, or Switzerland's US$17,010. Only three countries in the world are, in these terms, certifiably poorer than Ethiopia – Chad, Bhutan and Laos. There is associated with low GNP in Ethiopia a high infant mortality rate, low (6 per cent) access to clean drinking water, limited life expectancy (forty years), a very high illiteracy rate (the 1970 figure was 93 per cent, in 1986 perhaps 70 per cent), with about one doctor per 100,000 of population. Even before 1983, about 360,000 Ethiopian children under the age of four were reckoned to die annually. A country marked by such indices is always close to the margin. A drought which causes American or Australian farmers to tighten their belts may force upon Ethiopian farmers severe malnutrition, and worse.

Ethiopia is a highland, agrarian state. Land over 6,500 feet is basically free of malaria, and 70 per cent of the population of forty-two million is located at or above this line. Of the total population (encompassing both highlands and lowlands), 90 per cent is engaged in agriculture, which accounts for more than 90 per cent of Ethiopia's total foreign exchange earnings. But this population, before the present famine, was growing apace, expanding by between 2·5 and 2·8 per cent yearly. Food production fell far behind the rate of population increase.

The population increase is basically to be accounted for by the introduction of rudimentary health services. The increase in mouths to feed – over 200 persons per square kilometre in parts of Wollo, Tigray and Eritrea – led to more and more land being brought under the plough. This land, as a reader of Ricardo would expect, became increasingly marginal. Land that would normally have been left fallow was not. Hillsides that should have been left in the custody of natural cover were cleared. Yields declined in proportion to the increase in cultivated acreage. Holdings were unremittingly subdivided into smaller and smaller units, all of them over-farmed and over-grazed, the nutrients used up, the soil exhausted.

The effect of clearing the land and intensifying its use has proved, given the topography of northern Ethiopia, singularly destructive. The northern provinces are especially rugged. Small communities are cut from one another by deep gorges. Connecting roads are few and unreliable, and half the population in these craggy redoubts can be reached only by mule; in the rains (when rain comes), they cannot be reached at all.

When the cover is stripped from the steep slopes, erosion accelerates sharply. Ethiopia's forests have been cut down almost everywhere. They have been savaged for sawlogs, sold for various industrial purposes, burned for fuel. No more than 3·1 per cent of the entire country retains its forest cover; two decades ago, the figure was 16 per cent; a century ago, perhaps 44 per cent. In the north, no forests at all are left, and parts of it have become bare rock. With forest and other cover removed, the sun bakes then cracks the earth, turning acre upon acre into dust. With the onset of heavy rain, sharp gullies are cut and muddied waters sweep away an extraordinary volume of good soil. Today, probably about one and a half billion tons of Ethiopian topsoil are carried away yearly by action of wind and rain. (The figure given by the United Nations Food and Agriculture Organization two decades ago was one billion tons per year.) Because there is less and less wood, animal manure is increasingly pressed into service as fuel, thus depriving the remaining soils of nutrients which they require all the more because of intensified tillage.

The chief problem here seems at first to be population growth. The question that arises is to do with how such growth may be checked. The most ruthless means of checking it is famine, which brings misery, and anyway is purely temporary. Emergency food aid, on its own, avoids famine without encouraging growth and is not itself therefore a solution. It represents an important manifestation of humanitarian concern and is not to be disdained, but what it covertly points to is a more basic problem – that of development.

Development, as reflected in the GNP, does not necessarily eliminate physical misery and wanton destruction of human life. Many countries whose per capita income is relatively high may suffer high rates of infant mortality. Gabon had a 1982 per capita GNP of US$4,000, five and a half times that of Ghana, but Ghana had an infant mortality rate (1982) that was 9 per cent lower than that of Gabon. Development can be expected to bring with it lower death rates and, in consequence, lower birth rates. The short-term trend in all developing countries is for reproduction to shoot up as health facilities improve. As this improvement becomes tangible, and is so recognized by the populations affected, fear of early death among them declines and so does the rate of reproduction.

'It is highly probable that this extreme desire of having large families defeats its own purpose; and that the poverty and misery, which it occasions, cause fewer children to grow up to maturity than if the parents confined their attention to the rearing of a smaller number.' Thomas Malthus, much maligned as he has been, made this astute observation in 1798 in his *Essay on the Principle of Population*. But Adam Smith, in his essay *The Wealth of Nations* (1776), was more acute still. He estimated that, in north-western Europe, 'one-half the children born . . . die before the age of manhood'. He claimed it to be 'not uncommon . . . in the Highlands of Scotland for a mother who has borne twenty children not to have two alive'. He remarked that this 'great mortality . . . will everywhere be found chiefly among the children of the common people, who cannot afford to tend them with the same care as those of better station'. Smith's acuteness lay in his recognition of a logic, or compulsion: 'The poorest labourers . . . must . . . attempt to rear at least four children, in order that two may have an equal chance of living' (of reaching adulthood).

The levels and causes of infant mortality in Africa today are essentially the same as those which obtained at the turn of the century in the more developed countries. In Italy, at the close of the nineteenth century, infant mortality (per 1,000 live births) was more than 170 (as it was in

1982 in Malawi and Kampuchea). In France the rate was 160 (as in 1982 in Guinea and Yemen). In the United Kingdom, the rate was over 140 (as in 1982 Chad, Mauritania, Burundi and the Central African Republic). In turn-of-the-century USA, the rate was 120 (as in 1982 Sudan, Ivory Coast and India). In Hungary and some other Eastern European states, the rate was above 220 (for which in 1982 there was no parallel, according to the UN statistics, in any Third World state). The major infant killers in New York, Birmingham and Paris eighty-five years ago were those which now scissor their way through Addis Ababa, Nairobi, N'Djamena and the so-called Homelands of South Africa: diarrhoea; upper respiratory tract infections (especially pneumonia and bronchitis); and a great galaxy of infectious diseases, not the least among these being tuberculosis.

Although development, measured in GNP, does not directly reduce death rates, it does enhance the possibility of such reduction. Development – of transport, education, agriculture and the health services – is the key. It is to be presumed that this is the chief way, in Africa as elsewhere, that mortality will be cut, misery and insecurity reduced, and unbridled population growth checked.

In the developed world, it is obvious that rates of reproduction have declined sharply. In the developing states of Asia and Latin America, rates of reproduction have now slowed. Africa, the poorest of the continents, has reproduction rates which continue to soar. In Africa, the promotion of family planning is often regarded as a conspiracy of the rich and mostly white West against a poor and partly black South. Even where there is firm government support for family planning, as in Kenya, ordinary Africans are disposed to view such promotion with suspicion and cynicism. The population of the continent will move from 531 million in 1985 to 1 billion, on present trends, by the year 2000. But with development, we may expect this inordinate expansion to be checked. Without development and the long-term investment and aid which promote it, we can only expect continued population growth, consequential destruction of the environment and of other species, and the displacement, pauperization and immiseration of peoples.

Continuing population growth, the immiseration of peoples, the destruction of the environment and of other species, both plant and animal, are scything into Africa on a vast scale today. Ethiopia is the worst example. The country is poor and underdeveloped. Her circumstances render her acutely vulnerable. There are many factors which may diminish the surplus reserves available to Ethiopians in need, and one of the most important of these is weather.

In October 1984, nineteen of Ethiopia's twenty-six meteorological stations recorded a severe drop in rainfall. Six recorded rainfall at under half the expected rate. Twelve of the country's fourteen administrative regions were affected by drought. The northern provinces with a total population of over fourteen million were worst affected. Drought and famine have been an almost continuous and accelerating feature of life in these provinces since 1971–2. The famine of 1972–4 led to the overthrow of Emperor Haile Selassie. The penury of these years, meteorological and economic, was simply extended, at intervals, into 1984 and 1985. The number reckoned to have died in the famine of 1972–4 is put officially at 200,000; it may have been much higher – but the number swept away in 1984 alone is reckoned at about a million. And this, too, may be an underestimate. By 1985, the people of the northern provinces had experienced four crop failures in succession. In some parts of Wollo and Gondar, the rains had failed for as many as eleven years together. The failure of rain caused anguish. There are no well-developed means of channelling and storing water in Ethiopia. When rain did come, it was erratic, often thundering down, ripping rocks and boulders from the cleared hillsides, rolling them onto the flatlands below, increasing the difficulties of tillage and the farmer's burden of toil.

Drought does not directly cause famine, but it prepares the way for it. With the lack of rain in northern Ethiopia, rivers ceased to run. By March 1983, many people were having to cover as much as ten to twenty miles to find water. From Korem south to Dessie (the capital of Wollo) there are fifty-one rivers, thirty-four of which had dried up by 1984. The sparse rainfall was not unprecedented. In 1957–8, in Wollo and Tigray, there was drought, followed by locusts, followed by famine. In 1964–5, there was another drought, again associated with famine. In 1971–3 there was drought, not only in the north, but also in the east and south-east. The 200,000 death figure usually given for the ensuing famine covers only the north – specifically, Tigray, Wollo and north-eastern Shoa. These near-cyclical rain failures do not appear to have changed significantly, but the population has grown.

Ethiopia's north is bare. From 1982 to 1985, where there was planting, generally there were no crops. In cases where there were crops, there was a heightened incidence of pest infestation. Without water, there was no pasture. With neither water nor pasture, animals died. Those that did not die were slaughtered and eaten, or sold. The oxen are most important for the Ethiopian highland farmer – they draw his ancient, wooden plough. If the land is particularly difficult, he may require several oxen to do the job. But if he has weakened them, or worse, sold most of them, when the rain returns he can do nothing.

Families consumed surpluses from better times. Men and older boys began to range abroad in hopes of selling their labour – for cash or in some cases for food. Simply to give food to destitute people who have the strength to work may encourage a long-term enervating dependence. To remove these difficulties several food-for-work programmes were initiated. A labourer was given food in exchange for service – on irrigation, terracing or road-building projects. He then returned home. The disadvantage of such schemes has been that to carry bulky food improperly sealed over rough country delays distribution and induces spoilage, thus reducing the chances that it will reach those – the children – who need it most. Cash-for-work programmes have, on the whole, proved better; one important advantage is that they do encourage farmers in non-deficit areas to produce for the market.

The difficulty with all massive importation of free or cheap foreign food is that it undercuts incentives for local production and in the end destroys the home market for local produce. But free food, or food-for-work, or cash-for-work, were not available to the itinerant and increasingly desperate males seeking succour for themselves and their families during 1981–5. Some ranged as far afield as Assab, Massawa or Addis Ababa. In Addis they were usually shunted to the outskirts, held there, then transshipped mostly to the west and more fertile part of the country, irrespective of family or other connections. The prostitution of young girls became more common than ever.

The paucity of rain put pressure upon household reserves. These dwindling reserves forced able-bodied males to move to better-provisioned areas. The increased availability of labour in these areas lowered the level of wages generally. While wages and the price of animals dropped, the cost of grain sky-rocketed. In northern Wollo in 1984 oxen were selling for one eighth of their normal value; goats in Gamo Gofa sold for one tenth. In 1982, in Eritrea, the price of sorghum doubled. Similar developments were increasingly remarked throughout the north in relation to the cost of *teff* (the Ethiopian staple), wheat, barley and maize, where these items were available at all. Animals could be fed only at great cost; failed harvests mean scant fodder; so animals were exchanged at great loss to secure grain. The mothers and children and the aged at first waited in their home areas for help. But mostly their men could not get back to them, or had nothing they could bring even if they should return.

There are many first-hand accounts of these developments, in different areas, at different times, mostly private and unpublished. One of these, a report from a team of non-governmental relief workers in Wollo in late 1984, reads as follows:

In B— ... there are normally about 1,000 people, but now there are said to be 5,000, of whom 1,300 are camped near to us awaiting resettlement. This year due to the total failure of the long rains no teff ... has been harvested at all (their last crop was in September 1983) and no sorghum since January. There were two freak and savage storms in May and September, the latter damaging the crops that were still growing at that time. Ironically most cattle and goats look well fed; they are eating the dried remains of those crops which did grow. But for all that, they are producing little milk. Most farmers have cleared their fields now and have cultivated them ready for the short rains which ought to come soon (and the weather has become cooler and cloudier since we have been here). But few, if any, have seeds to sow, having eaten their seeds as well as their food grain. Many people have been reduced to such a level of desperation that they have been dismantling their houses to sell off the poles and the thatch to buy any little food on the market. And the market square always has a number of the very hungry picking up individual spilt grains.

Since many people have sold their buildings unknown numbers of people are living around the town on the dry rocky hillsides and barren fields in small bivouacs of twigs and branches. The extent of this bush settlement is only apparent at night when the hillsides flicker with little fires.

Access to B— on the all-weather dirt road has been improved by local food-for-work teams, but we keep puncturing landrover tyres on long acacia thorns. The road to H— remains too difficult for trucks. For this reason it has been necessary to move supplementary food there by camel train. At the weekend we used 90 to carry about 6 MT – a low per camel capacity, because they too are moderately to severely malnourished.

The general situation here is manifestly poor, and the immediate future is rather bleak. There are many thin and destitute adults and children of all ages. Some are clearly very desperate. But everybody is doing their utmost to assist us in our work. We have had excellent cooperation from the local administrations at all levels. We also have a very good working relationship with the administrator here in B—, who is most concerned both for our work and for our welfare.

Families ate their grain, of course, and then, when that was finished, they ate any seed they had stored underground, intended for planting when the rains returned. There was no let-up in the drought, and there was little access to government stores. The government, however diligent and resolved, was itself only a corporate pauper: there really were no reserves to speak of. And in any event it was caught up in civil war, which swelled the size of the army from 40,000 to 300,000 men in less than a decade.

People exhausted their grain, then disposed of their animals, then spent their money. They borrowed from neighbours and relations. Where they could, they sold their clothes, their furniture, their amber beads, their silver jewellery. They began to eat coffee leaves, wild grass seeds and cactus fruit. Finally, they let foot follow foot down the road to the burgeoning shelters and distribution centres that increasingly dotted the countryside.

It is not clear that there simply was no food. It seems likely that in 1972–4 there was sufficient to serve the purpose, but that it remained undistributed. It is beyond question that, by 1983–4, the shortfall was far more severe. The production of food in Ethiopia (as in Africa generally) had declined seriously over 1981, 1982 and 1983. On 6 November 1984, Kurt Jansson was appointed coordinator of the UN emergency relief effort in Ethiopia. By early December 1984 he was installed in the Economic Commission for Africa in Addis Ababa. Following a survey completed by a Food and Agriculture team, Jansson was able to report to his headquarters in New York that the total harvest from the long rains of July–September (the *Meher*) would not exceed 5·6 million tons. This represented a deficit of 30 per cent even in relation to the three drought years that had just gone before. The sharpest falls were in the production of maize, *durrah*, pulses and *teff*. The shortfall was estimated as perhaps in the area of two million tons, an amount that could feed as many as eight million people.

There could be no question of the dimensions and significance of the Ethiopian crisis. Awareness of the problem exacerbated it. Those who held stocks surplus to their needs either genuinely feared that they would not remain 'surplus' or correctly anticipated that buyers could be induced to exchange money for stock at several times the normal price. In 1972–4, food surplus areas were adjacent to food deficit areas; 'surplus' food was available even in food deficit areas – but always at a higher-than-normal cost. In 1982–4, the situation was in essence the same. The numbers, however, had significantly changed. In absolute terms there was much less food available in relation to the number of people who needed it. Many more people were affected; many more of those whom their neighbours were accustomed to regard as 'rich' had become poor.

Government agricultural policy, however well intentioned, was unsuccessful. Following the *coup* in 1974, the new government moved to implement agricultural policies, which proved popular. In 1975, farmland previously owned by the aristocracy was nationalized and placed in the hands of smallholders, often occupying no more than two hectares, and never more than ten hectares per family (about twenty-five acres). The small farmer was then in control, but had no security of tenure. The newly introduced farmer associations decided upon the allocation of land. The government took it upon itself to market peasant produce, for which it paid little, on occasion less than cost (to favour city-dwellers, the army and the bureaucracy). This further encouraged black-marketeering, which meant further governmental interference. Food production very nearly stood still. The government's post-revolutionary redistribution had the twin effects of encouraging further population growth and discouraging greater agricultural output. In 1978 the government made state farms a top priority, effectively tripling their acreage over the next eight years (to 6 per cent of total crop production). It also accorded high priority to production cooperatives (about 1 per cent of farmland under cultivation), intending that, by 1994, these should absorb as much as 50 per cent of farmland. Both cooperatives and state farms were accorded significant percentages of Ethiopia's thin surplus – but state farm production declined, while the cooperatives proved unpopular.

Given insufficient supplies, non-existent roads, lack of transport and lack of security in the rural areas, people began to move to relief centres, like Korem. The idea was not that people should settle down at such centres, but that they should come, collect relief supplies on a fortnightly or monthly basis and then return home. But 'home' might be fifty or sixty miles away, to be reached only on foot, and there was perhaps less certainty of resuming a 'normal' life on return than on first departure.

If people were attracted to these relief centres, it was because they had little choice. The consequences of staying at home with little or nothing to eat were all too obvious. But the centres, which began to spring up everywhere, were often a good distance from where they lived.

The services they provided were rudimentary. Though called 'shelters', there was in fact little shelter available. Up to the beginning of 1985 the centres had only limited reserves of food. Not only were people still starving to death, they were also dying from exposure. There was little or no clean drinking water. Respiratory and other infections associated with crowded and insanitary conditions began to take their toll. With the onset of rain, however limited its duration, people wanted to go home, and were encouraged to do this: the idea of permanent communities of beggars appealed neither to governments nor to private donors, Ethiopian or other.

The affected people were mostly rather old, or very young, or women. Able-bodied males, or at the least young males, were likely to be pressed into the army, or resettled to the west or south to clear yet more forest. The old tended to think that they would die in these centres. Mothers thought that at least their infants might be saved. It did not require much rain to encourage famine victims to quit the shelters (as at Ibnat in Gondar Province in November 1984), despite the considerable distance which might separate victims from their home areas, uncertainty about the weather they would find there, and lack of seed to plant and oxen for ploughing.

The number of relief centres increased from 120 in 1983 to 225 at the end of 1984. By mid-1985, there were twenty-three major shelters, 162 feeding stations and 286 food distribution centres. Over 150,000 orphans were being cared for in the centres. The number of people affected by famine has been variably estimated at around five million in 1983, over six million in 1984 and well over ten million (approaching eleven million) in 1985. As for the numbers of persons receiving aid in any form, the Ethiopian government gave a figure in excess of seven million. In May 1985, *Le Monde* claimed the correct figure to be 30 per cent less. American officials claimed at about the same time that 90 per cent of relief aid was reaching those for whom it was intended.

The Ethiopian government, through its relief agency (RRC), had been warning foreign aid donors of the growing famine since 1983, warnings repeated in February, March and April 1984.

At a meeting held on 17 September 1984, the association coordinating relief work by private organizations in Ethiopia (CRDA) at last felt constrained to take the unprecedented step of telexing its own alert to the international community (governments, the UN and private donors) in the following terms:

This appeal arises out of a meeting convened by the Christian Relief and Development Association (which is a coordinating and funding Association consisting of twenty-six churches and voluntary agencies) and attended by representatives of the undersigned churches and agencies working in relief throughout Ethiopia.

The purpose of this telex is to express our deep concern at the gravity of the famine here in Ethiopia and at the desperate shortage of relief food, and to request immediate and extraordinary action by all relief donors to meet the crisis.

Ethiopia has not experienced a food shortage of this magnitude within living memory. In terms of geographical extent and population affected, it vastly exceeds in severity the drought and famine of 1973, when three Regions were affected. Today twelve of the fourteen Regions are affected by drought and death by starvation is occurring in six of these. More than six million people are estimated to be affected by food shortage.

The number of people arriving at feeding centres far exceeds the supplies of available food. Currrent estimates made from World Food Programme reports indicate that only 100,000 tonnes of relief supplies are scheduled to arrive by the end of December 1984. This is sufficient to feed the affected population for only thirty days. Moreover, extensive crop failure in all parts of the country indicates that not less that 60,000 tonnes of relief food will be required per month until December 1985. There is no doubt that if substantial quantities of food are not forthcoming immediately, hundreds of thousands of people will die. This can be avoided.

We are aware of the logistical and bureaucratic constraints but we are confident that through a concerted effort by the Government, International Bodies and Voluntary Agencies, these difficulties can be overcome.

In view of the abundance of food and the financial resources in many countries we implore all donor nations and agencies to give urgent consideration to:

1. Increase food supplies to Ethiopia in the current year, by whatever means at their disposal.
2. Ensure that adequate grain supplies reach Ethiopia during 1985, which are estimated at 650,000 tonnes.
3. Consider the allocation of financial resources to meet in-country transportation and distribution expenses including the provision of trucks and spare parts.

We are taking the unusual step of contacting Governments, the UN and donor agencies directly out of the conviction that only immediate and massive action can arrest this famine.

There was little apparent government reaction to this startling message. At about the same time, a BBC television team managed to get into Addis Ababa, and up-country into the camps, including Korem. There had been reports on this famine in the press for the two previous years.

But when the film of devastated famine victims was shown in Britain in October 1984, the popular reaction in Europe, the United States and later the world was overwhelming. In turn, this reaction, among other effects, dissolved somewhat the rock-hard suspicion of the Ethiopian government towards further outside news coverage of developments there.

# 4  Aid: The Politics of the Pendulum

In 1984 a group of British pop stars led by Bob Geldof came together and styled themselves Band Aid. To help victims of the famine in Ethiopia, they recorded and released what was to become a hit single, 'Do They Know It's Christmas?'. The amount that this title earned in Britain was not expected to exceed £4m. (about US$5·6m.). Worldwide, it was expected to generate about £8m. (US$11·2m.).

In January 1985, following the success of Band Aid, forty-five American pop artists called up by Harry Belafonte set out on the same path, recording 'We Are the World'. The song was written by Michael Jackson and Lionel Ritchie and arranged by Quincy Jones. Dollars were not slow in coming, and the total sum earned has usually been given as US$50m.

In late March, Canadian pop stars joined this procession of well-intentioned pipers. They saddled themselves upon a tune, 'Tears Are Not Enough' (penned by Bryan Adams), ran it as a single, and in under two weeks brought in over a quarter of a million dollars.

Julio Iglesias was one of about half a hundred Latin pop stars who forgathered in April to record 'Cantare, Cantaras' in Spanish, Portuguese and English, with rumoured sales of as many as ten million copies; similar recordings were made in France and elsewhere. By mid-July 1985, Bob Geldof, mildly delirious with his earlier success, swung into action again. This time he organized a Live Aid rock telecast, a marathon extravaganza. Beamed into the heavens simultaneously from Philadelphia and London, it bounced back across the world. It is said to have pulled in about US$50m.

Pop stars, established charities and ordinary people were making a singular effort to do what they could to diminish the pain endured by so many elsewhere. By February 1985, about US$70m. had been collected in America to fight African famine. By June 1985, these private contributions had swollen to US$120m. In the United Kingdom, the parallel amount privately donated was in the area of £67m. (US$93·8m.), an outlay 300 per cent higher per capita than in the USA. In Australia, aid privately donated to Africa (over the period July 1984–June 1985) amounted to AU$18m. (US$12m.), 60 per cent higher per capita than the USA figure. Europe and Canada and Japan realized similar per capita totals. This private fund-raising activity was extraordinary and, in its global reach, without parallel.

The sums generated were substantial, but this was not their most significant feature. This private outflow of sympathy and material support had the effect of placing pressure upon *governments* in wealthy democracies to do much more to help. The bulk of all emergency aid was Western, and most governments felt it prudent to step up levels of such aid, first and foremost to Ethiopia. The Reagan administration appeared to drag its feet on aid to Ethiopia for ideological reasons, but within weeks of October 1984 it suddenly released sums many times greater than previous emergency outlays. The Italian government, in the wake of widespread public concern over famine, moved to double the level of its aid programme. British emergency aid to Ethiopia, at government level, multiplied to £44m. (US$61·6m.) in the nine months following the October BBC broadcasts. This was an unusual case in that the sum donated by government was less than the amount raised privately. The government, accordingly, was roundly attacked by both the Labour and Alliance opposition for being niggardly. Australian government aid to Ethiopia expanded from AU$4·3m. in 1982–3 to AU$27m. (US$18m.). All told, within six months of the October 1984 televised coverage of the Ethiopian famine, as much as US$2b. had been pledged for aid purposes by various governments. Western donors subsequently earmarked nearly $3b. for African relief in 1985, the US government covering about half of this amount.

The upshot was that food supplies were increased in sufficient quantity to check and reduce the tidal wave of deaths. Grisly televised coverage had swept away, at least temporarily, the indifference of most foreign viewers. Private aid accumulated. Governments joined in and the volume of their offerings easily overtook that of the voluntary agencies. The bulk of such aid was food, off-loaded at Assab, Massawa and Djibouti, which in the end lacked the capacity to cope. Ships had to wait at anchor. Cargoes off-loaded had to wait in turn to be cleared from docks. Some shipments, as at Assab, had to wait in the sun or rain; they burst, fermented and rotted.

Even as the human misery was attenuated, the underlying causes of misery remained quite untouched. Sometimes there were no roads, or 'roads' were impassable; the capacity of rail systems was limited, and these too became inoperable; the trucks were old, and axles cracked in these conditions; so planes were used, smashing their tyres at $5,000 per time; air-drops were little more than a gesture. Had the country been reasonably well developed, there would have been no significant transport problems, but she was not. Her most violent haemorrhaging was stanched by emergency aid. But the essential need was for the sort of development which would obviate the necessity for such deliveries. The point, if made once, was made a thousand times – at the UN, by private donors, by aware government officials, both inside and outside Africa. But nothing (or next to nothing) was done. If anything, development aid went backwards – to the point where it proved appropriate to speak of the development of underdevelopment.

Emergency aid naturally was excessively concentrated upon 'emergencies'. To develop countries more fully removes the need for much of that sort of help. The trouble is that food aid, for Western government donors, is especially easy to give, even though it may be given slowly. There are mountains of food stockpiled in Europe, America, Canada and Australia, production encouraged by substantial government subsidies to farmers. Giving the food away, or selling it at less than market price, is one way of disposing of a domestic embarrassment, and it makes a good impression. But except in the very short term, it resolves nothing, and probably ensures that the same difficulties will later re-emerge with redoubled fury.

Maurice Strong, Executive Director of the UN Office for Emergency Operations in Africa, declared (in June 1985) that US$1·5b. in emergency and rehabilitation assistance was needed for Africa in 1985. He stated that food aid was 'close' to the targeted figure, but that other types of aid, specifically for agricultural production, were lagging behind. 'What worries me,' he said, 'is that the Western world may find it much easier and simpler to provide food aid and relief needs rather than the long-term technical and capital assistance' required. In view of the paucity of aid to the world's thirty-six poorest nations, UN planners devised for the Conference on Trade and Development a programme to accelerate aid by 66 per cent over the years 1981–5, compared to 1976–80. By the end of 1983 a great space had opened up between hope and achievement, the increase being only 15 per cent, and per capita incomes subsequently continued to decline. In 1984, there was a drop of $300m. in US multilateral assistance to sub-Saharan Africa for development purposes. The Labour leader of Britain's Opposition, Neil Kinnock, claimed in the House of Commons in mid-July 1985 that the Thatcher government had made a real cut of $40m. in aid, ten times the amount he thought likely to be raised by the Live Aid concert in Britain.

Even the amount of emergency aid provided by governments seemed to invite controversy. At their December 1984 summit in Dublin, the ten leaders of the EEC pledged 1·2m. tons of grain for Africa. This and other measures were taken to signal a new commitment to assist in the continent's development. By the middle of 1985, however, a detailed report by the British House of Commons Foreign Affairs Committee revealed that the EEC was operating a system of 'double counting'. According to the committee, the EEC's restricted commitment to supply Africa with 500,000 tons of food was made to appear twice as high. Even Britain's emergency transfer in 1984 to Sudan of £22m. and to Ethiopia of £44m. was achieved only by suspending commitments to other needy states, so as not to exceed the existing aid budget of £89m.

There was a dramatic contrast between the foot-dragging of rich governments and the readiness of their citizenry to come to the assistance of desperate people in need. In May 1981, Ethiopia's Relief and Rehabilitation Commission (RRC) called attention to drought, predicted famine and sought emergency assistance in the form of grain, oil, vaccines, transport and the like. The same warnings were repeated in 1982. In 1983, the figure given for the drought-affected population approached four million. In August 1984, the RRC announced that no grain had been in stock since mid-July. In October 1984, the RRC figures given for the drought-affected leapt to 6·4m., and in December they rose by a further 20 per cent to 7·7m.

The RRC was originally established in 1974. It had a very troubled beginning, troubles far from over in 1983, by which time it had accumulated thousands of employees, hundreds of distributive outlets, shelters, numerous feeding centres and stores, an airline and a thousand vehicles. But despite all its personnel and property, the RRC was a mouse atop a mountain of need. Its estimates could not be regarded as statistically reliable, but they painted a reasonable picture of the unravelling of a catastrophe. Unfortunately, not much notice was taken of it. The RRC was not entirely ignored, or entirely refused. But, as the RRC's Chief Commissioner claimed in October 1984, the pledges made earlier in the year 'were nowhere near our requirements'. After October, in the wake of popular support for Ethiopia, especially in the West, Mr Dawit Wolde Giorgis protested that the famine 'could so easily have been avoided' if due attention had been paid to RRC warnings by donor governments.

Up to 1974, Ethiopia's main trading partners and allies were the USA and the EEC. With Ethiopia's entanglement with the Soviets, development aid from the West was slowed down; in the case of the USA, it was shut down. By March 1983, when the Ethiopian crisis was approaching its most critical phase, US development aid had completely dematerialized. There was some US aid, of course – emergency assistance and the charitable work of US-based non-government organizations such as Catholic Relief Services (CRS) and World Vision. These organizations themselves often served as private conduits for public (governmental) assistance. But even here there was significant opposition within the US Agency for International Development (AID) to supplying private charities with humanitarian aid for the Ethiopians. The grounds for opposing humanitarian aid were (among others) that it would allow Ethiopia to redirect, from food to war *matériel*, the money it saved, thereby more readily containing rebel and secessionist elements, especially in the north.

The US State Department rejected the idea that the aid pendulum was in any way swung by political considerations. In June 1983, the Department reported that US$4·8m. in food aid had been supplied to Ethiopians, in rebel and non-rebel areas. In September, the Department announced further aid in the order of US$800 thousand through the UN Disaster Relief Organization (UNDRO) for emergency transport. According to a press release from the US embassy in Addis Ababa, total US emergency aid to Ethiopia for the fiscal year 1983–4 (1 October to 30 September) amounted to nearly $17m. The RRC complained in October 1984 that such assistance was inadequate for the crisis Ethiopia confronted.

As early as 5 May 1983, the United States government had officially concluded that 'a disaster situation' of mammoth proportions already existed in Ethiopia. But there was no significant increase in emergency aid. It was only after the extent of the Ethiopian famine was publicized in the media from late October 1984 that action was taken. By this time, tens of thousands had died who could have been saved. President Reagan, under pressure, overcame his hesitancy, and within two months (in January 1985) had authorized disbursements almost fourteen times

($235m.) greater than the $17m. conceded for 1983–4. Official America's unwillingness to help starving Ethiopians was seen by some as a punishment to the government under which they lived. It was only the post-October 1984 explosion of public concern that turned Mr Reagan's government round. That is not to say that the RRC represented a model of correct behaviour. In December 1984, when the supply of emergency food aid was dramatically improving, the RRC understated the volume of aid coming from American sources by 77 per cent by the simple expedient of omitting, in its full-scale, end-of-year *Review of the Current Drought Situation in Ethiopia*, all mention of US non-governmental aid. Addis Ababa's judgement was just as clouded by resentment as was Washington's humanity by its cold obsession with the maintenance of strategic superiority over the Soviets.

The State Department's denials in 1983 of political considerations in the distribution of emergency aid were transparently untenable. The termination of development aid to Ethiopia, ostensibly because of her nationalization of US assets, meant that US emergency aid, up to April 1985, could not be used in such a way as to support road or well construction, irrigation projects, the purchase of trucks or seeds, or any other activity that could be construed as 'developmental' – as involving anything more than just keeping people alive. USAID, apart from severely restricting the use of humanitarian assistance where this could be said to have any developmental effect, took considerably longer than normal to process Ethiopian emergency food requests that first began to reach USAID from non-governmental organizations (principally CRS and World Vision) in December 1982. This was the conclusion of a report prepared at the request of Representative Byron Dorgan by the US General Accounting Office (GAO) and published in the second quarter of 1985. Approval of these 1982 and 1983 requests for emergency food aid to Ethiopia took about half a year to clear. The GAO explained this delay (in part) thus: 'A continuing concern of AID and the Department of State is whether food is used to support an Ethiopian government that is openly hostile to the United States.'

The paucity of Western government aid to Ethiopia in the pre-October 1984 period was sometimes explained and justified or excused on the grounds that Ethiopia's RRC had exaggerated the number of persons affected by famine. In fairness, it must be said that it is extremely difficult to ensure that any figures in such a context are truly reliable. At the same time, it seems equally clear that the effort of Ethiopia's military government to set up a system providing early warning of impending food shortages was far from perfunctory. Their attempt to do this stemmed directly from the devastating drought and famine of 1972–4 in the northern highlands. It was the failure of the Emperor's U S-backed government to cope, especially with the 1972–4 famine, that led – more than anything else – to the rise of the Colonels in the first place. The military, who were openly in charge by October 1974, governed through an unwieldy committee of 120 men, expressed the radical views consistent with their unstable and exciting circumstances, and were received in the official circles of the West with a combination of uncertain disbelief and calculated unease – and nowhere more so than in the United States.

The military government that took power in Ethiopia declared the country to be a socialist state. It nationalized various companies, clamped down on trade unions, and (in 1975) also nationalized land, both rural and urban. Haile Selassie's *laisser faire* policy seemed to have provoked the opposing extreme of an entirely state-controlled economy. These nationalizations, however well-intentioned, brought little in the way of viable solutions. Smallholder farm production did not increase, savings declined and the state farms were by and large unsuccessful. By January 1983 Ethiopia's leaders were already backpedalling on their root-and-branch nationalization policy. They promulgated a new joint-venture code designed to attract capital frightened off in 1975 and after. From 1983, encouragement of private-sector investment, local and foreign, slowly accelerated. In early August 1985, Ethiopia's Foreign Minister, Goshu Wolde, spent a week in Japan. There he promised Tokyo (and thus the West) further revision of the laws governing foreign investment in Ethiopia, and also voiced his government's desire to secure significant private Japanese investment.

Dire circumstances, such as those of 1972–4, seem to conjure up correspondingly stark solutions. The nationalizations, if they solved little, symbolized a lot: the surging expression of an uncompromising desire to break with all those procedures and circumstances which were thought to have generated the miseries of the past. The value of the U S assets that had been nationalized was risible (U S\$22m.) but this did not stop the American government from using these developments as a bar to future good relations. Of course, just because the foreign assets taken over were insignificant, it would have been better, objectively, for Ethiopia to let them be. Ethiopia had, and has, so little that the priority concern cannot really be to appropriate what there is, but to build, in cooperation with the international community as far as possible, what does not yet exist.

From as far back as 1946, Ethiopia's major source of military and other external aid had been the USA. She remained the chief recipient of US aid in Africa for almost three decades. But Egypt, after the 1972 expulsion of Soviet advisers and equipment, and after a daring surprise attack on Israeli defences across the Suez Canal in the October War of 1973, was rewarded for her resilience by the restoration of diplomatic ties with the USA, the Sadat Presidency receiving the personal embrace of Richard Milhous Nixon in the June heat of Cairo in 1974. These developments paved the way for an initial aid package to Egypt of $250m. for the fiscal year 1974–5 alone. The emergence of Egypt as a new American client state along the Red Sea, given the close relationship between Egypt and Sudan, critically undercut Ethiopia's appeal as a recipient of US aid in this vital strategic zone. The fact that, in 1974, control of Ethiopian affairs had passed to a radical team in no way helped Ethiopia's position.

The old regime of the Emperor had died with the tens of thousands who starved to death in 1972–4; the figure usually given is 200,000 but estimates stretch to 400,000 and more. Public discontent was directed against the old regime, which had lasted since the earliest years of the century, heroic in its resistance to Italian imperialism in 1935 and 1936, but now tainted by its impotence and indifference towards the extinction of so many. The USA, given the new possibilities offered by Egypt and Sudan, together with other factors – not least, those relating to the development of sophisticated satellite surveillance – had less need by 1974 for her always tetchy (but now restive) Ethiopian partner. Ethiopia, given her domestic, revolutionary drama, was ideologically less able and disposed to seek help from America just when help was most needed. The studied and undiplomatic hostility of the USA (under cover of legitimate dismay at obvious human rights violations) towards the blood-smeared fumbling of the new men in Addis Ababa seemed calculated to drive them further left, to alienate from them traditionalist and wealthy elements, as also from one another, and so cause the self-primed demolition of the junta.

The Arab states across the Red Sea counted on the collapse of the new regime. For a time, so did Nimeiri's Sudan. Neighbouring Somalia not only expected Ethiopia's collapse: it staked its future on it (deftly angling for support from the Carter government in the process). The new order, however, was to survive – bloodily, cruelly, but survive all the same. There is little evidence of any significant and intelligent activity by the USA to save Ethiopia from the quagmire into which she was sinking. Voice of America broadcasts beamed back in Amharic found fault at every turn. US foreign policy assumed a self-righteous mien. Arab allies across the Red Sea wanted Eritrea to be detached from Ethiopia. Somalia wanted the Ogaden. It would be so much simpler to allow Ethiopia to disintegrate. It was difficult to assume that in 1977 the Soviets would or could pick up the pieces, or stay the course if they did. The Somalis would be rewarded with land; it could be expected that the Soviets in Somalia (as in Egypt) would again be compelled to pack their bags. It would have appeared that the elimination of the Soviet presence from the entire area of the Red Sea and the Horn could not be bought more cheaply than by allowing Ethiopia, after 1974, to dissolve. In a way, it was convenient that the Emperor had been retired. Much easier to blame the new boys for the débâcle. No significant attempt was made to bring peace to the region, to dampen the fires of war; the effect, increasingly, was to demonstrate that socialist regimes are untenable. There was no complete rupture of US–Ethiopian ties, only a slow, sure decline.

The Ethiopian military came to power in the form of the PMAC – the Provisional Military Administrative Council. It was called the Dergue for short – Amharic for 'committee'. The Dergue acted energetically, its object being to satisfy the requirements of an unsettled population. It tried to check famine and erosion and to correct the distributive abuses of the previous regime. The initial measures adopted in 1974 and 1975 pleased many and alienated others, not just outside, but also inside, the Dergue – a far from stable entity. Lieut.-General Aman Andom, who followed Haile Selassie as substitute Head of State, was terminated with extreme prejudice in November 1974. It is claimed that he did not order sufficient troops into Eritrea (he was himself Eritrean) to counter the secessionists. Dozens of other officials, among them seventeen generals, were summarily executed at the same time. Brigadier-General Teferi Banti assumed formal control. Increasing opposition to the Dergue was met with arrests and executions. Major Sissay Habte, the Dergue's No. 3, along with several others, was executed in July 1976 for alleged counter-revolutionary activity. In February 1977, General Teferi Banti himself (among others) met the same fate. Lieut.-Colonel

Mengistu now assumed the purple, losing only a few months in arranging or accommodating the execution of his Vice-Chairman for trafficking with the Ethiopian People's Revolutionary Party. The EPRP opposed both continued military government and Ethiopian control of the province of Eritrea, and was the most important party on the left. A fully fledged civil war finally erupted between the Dergue and the EPRP in late 1976.

The EPRP embarked upon a campaign of urban terrorism, targeting government officials and sympathizers, thinking thereby to weaken support for the official line within the Dergue. The effect was precisely the reverse. Although the Dergue did not develop its all-out campaign of 'red terror' straight away, it gradually eliminated those of its members who might be suspected of sympathy towards EPRP objectives. In this way, the ground was prepared for the effective elimination of the EPRP opposition, an objective finally secured by January 1978. A tit-for-tat civil war had gathered momentum, through 1977, reaching a crescendo of street shoot-outs, executions, round-ups, house-to-house searches and massacres, with a final toll of thousands dead.

The extinguishing of significant internal opposition to the Dergue was bought at a terrible price. It was far from clear, at the beginning, how the contest would be resolved. Victimization was widespread, domestic tragedies unnumbered and the appearance of chaos limitless. In 1976 and 1977 one could still fly into Addis Ababa, as under Haile Selassie, without a visa – which even in the most peaceful of times could never be done in Tanzania or Kenya or Ghana or Madagascar. But at night, in the eery quiet, hotel guests were rocked to sleep (or wakened from it) by the patter of gunfire. It echoed below throughout the city until dawn, heralding an intermission from killing; then after dark the fearsome but idiotic play of lead on flesh would begin again. In this play, the Dergue, with Mengistu at its head, proved as ruthless as those who flew the flag of resistance. Mengistu proved himself a master of cold, quiet cunning (very much in the manner of Haile Selassie), and was not without a good deal of luck.

Ethiopia had always been governed by imperial *fiat*. The transition from the imperial regime to a new dispensation would have been troubled even in the most auspicious of circumstances. The Dergue, as a governing committee, was not a viable institution; it could work only if the entire country agreed on public policy, and such agreement was not forthcoming. The members of the Dergue would consult and be consulted by outside elements. Its members, as a matter of course, would play their cards close to the chest: mutual suspicion was the order of the day. It led to civil war. This, in turn, encouraged both secession, and a Somali lunge for territorial advantage. The first priority of the Dergue, in removing the Emperor, was to attend to problems of drought, famine and development with which the old regime had failed to cope. Given divisions within the Dergue, the new first-priority problem speedily and unceremoniously became that of establishing and retaining authority. The Dergue, having given birth most painfully to the leadership of Colonel Mengistu, was required to defend Ethiopia both against secession and foreign invasion. This superseded all other objectives.

The Dergue's most pressing difficulty – apart from survival – became Eritrea. Eritrea had been reluctantly absorbed as the fourteenth province of imperial Ethiopia in 1962. It had been federated with Ethiopia in 1952 under UN auspices. Eritrea was fairly democratic, Haile Selassie's Ethiopia was coolly autocratic. Either Eritrea would subvert autocracy, or autocracy would swamp Eritrea. It is the latter course which was pointed to by the events of 1962. The Emperor's takeover was reviled by many, but certainly not least by lowland Muslims opposed to domination by highland Christians. Nor was the Emperor's move to abolish federation unexpected. When Egypt's President Nasser was at daggers drawn with Israel and with the West, the Eritrean Liberation Front (ELF) was provided with office space in Cairo, and in the year before the expected eclipse of the Eritrean–Ethiopian federation, the ELF committed itself to the cause of armed struggle. The Eritreans initially operated from Sudanese bases and made a speciality of small-scale attacks on lines of communications and security – rail, roads, bridges, depots, police posts – with occasional hijackings and assassinations. In November 1970 the commander of the Ethiopian army in Eritrea was killed in an ambush on the Asmara–Keren road. In September 1973 the Ethiopian deputy-commander of the Second Division was killed in another ambush, this time in Asmara itself. As the hold of the Emperor conspicuously slipped, not only did his army begin to stir, but so did everyone else: the Oromo (Gabbra or Galla) in the south, the Somalis in the east and the Eritreans most importantly in the north. Starting in 1975, the revolt in Tigray province, on Eritrea's southern border, was to complicate matters still further.

The removal of the Emperor could, theoretically, have had two quite different effects: it could have raised Eritrean expectations that a new government in Addis Ababa would provide the best hope of a sensible deal; or it could have spurred on Eritrean secessionists in the thought that there could be no better time to strike the fetters of Empire than when that Empire was itself pinioned by the intractable forces of domestic revolution. This second effect proved the more potent. There were major clashes among the variegated Eritreans themselves (Christian, Muslim; Tigrinya, Danakil; highland, lowland). Just as the Dergue physically liquidated opposing members, so did the Eritrean liberation groups, with violence and cruelty. The Eritreans and the Dergue became trapped in a dialogue of the deaf: the Eritreans would cooperate with the Dergue if the latter would make a prior promise of Eritrean independence; the Dergue would allow full regional autonomy if the Eritreans accepted that the territorial integrity of Ethiopia was not to be challenged. The Eritreans argued that they were fully entitled to self-determination. The Dergue argued that Eritrea was one of the most historic parts of Ethiopia, that it was in no worse state than any other province and that there was involved in this secessionist project an (Arab-orchestrated) attempt to deprive Ethiopia of the only two outlets (Assab and Massawa) to the sea directly controlled by Addis Ababa.

While Ethiopia was wracked by the revolutionary developments of 1974–5, the three factions of the Eritrean liberation movement, goaded into a mutual embrace by Arab backers, cooperated in an all-out attempt to terminate the Ethiopian presence in the north. The Ethiopian army was worsted in an entire succession of engagements and in February 1975 came close to losing Asmara, the provincial capital. Washington was slow to respond to an urgent Dergue request for military help. None the less, US military assistance doubled in 1975 *vis-à-vis* 1974, and was higher again in mid-1976. The US position was hesitant throughout. There was no simple desire merely to break relations with the Dergue. But domestic Ethiopian policy (nationalizations for example), the fear of being tainted by that policy (especially in regard to summary executions without trial), the wish to punish the regime for courting the Soviet bear, the lessened dependence on Addis Ababa anyway, all encouraged the USA in its hesitancy. But by the close of 1976 a clear direction was finally set. President Ford called a halt to military grant aid. After Colonel Mengistu destroyed his remaining adversaries within the Dergue in February of 1977, the Carter administration suspended all arms supplies. In April it announced a reduction in the level of military advisers, and declared that the Kagnew communications base would be phased out. Mengistu, in response, precipitately shut down Kagnew and all US Information Service activity, and expelled the US military advisors.

The Eritrean challenge to Ethiopian territorial integrity was by no means the only one, or even, as it happened, the most important. The fact that the Ethiopian turmoil had encouraged the Eritreans, and then the Tigrayans, to strike gave encouragement to Somali guerrillas in the Ethiopian Ogaden, as well as to their backers in the Somali capital of Mogadishu.

Somalia, independent in 1960, represented a merger of former British and Italian littoral colonies. Its constitution contained a commitment to the 'recuperation' of Somali folk located in French Somaliland (Djibouti), Kenya and Ethiopia. This was tantamount to a commitment to war, and war there had been, up to 1967. A guerrilla campaign had been conducted against neighbouring Ethiopia and Kenya, until an uneasy settlement was reached through the mediation of Zambia's Kenneth Kaunda.

The Somali colonies of Britain and Italy, as well as of France, had served a strategic, not an economic, purpose. To make a new Somali republic financially independent was to prove a far more daunting task than that confronting most African territories. The ethnic tensions (inside Somalia, they were usually condemned as 'tribal') that had been directed outwards during the confrontations with neighbours that followed independence were turned upon Somalia itself after the peace of 1967. Out of the military *coup* of 20 October 1969, Major-General Mohammed Siad Barre emerged as the new Somali President. By 1973, the concern with 'reclaiming' what was regarded as Somali territory again came to the fore. Somalia had in fact for a long time given considerable support to the Western Somali Liberation Front (WSLF) and by 1975 was also supporting the Oromo Liberation Front (OLF), both operating inside Ethiopia. Somalia was sufficiently confident of its ground, and of the Dergue's vulnerability, to launch its own army in a massive operation across the border in July 1977, mauling Kenyan forces in the process. Barre's decision was made easier by Carter's suspension of arms supplies to Mengistu in February. It was essential that the Somali leader acted quickly, in advance of any Soviet move to replace the Americans as armourers to the Dergue. Barre miscalculated. His hasty expulsion of the Soviets from Somalia in November 1977 would not reverse the humiliating retreat that was later to be forced upon him.

During the period 1974–7, the Ethiopian military were faced with a chilling series of challenges: revolutionary ferment in the country as a whole; the divisions within the Dergue; a murderous civil war centred on the capital city itself; several regional uprisings and attempts at secession; and finally, a full, armed assault upon the country's southern marches by a hostile and disciplined neighbour.

These difficulties inevitably distracted from the more important problem of attending to the destitute, of promoting terracing, water conservation and the like. A cold, rough justice was very much the order of the day. Poor and uneducated soldiers suddenly saddled with power do not make indulgent policemen. The Ethiopian Relief and Rehabilitation Commission was none the less set up. The RRC was not in an ideal position to do its job because it had very little detailed, reliable information on the extent of Ethiopia's crisis, and very limited reserves to distribute in any case. An Early Warning System (EWS) against famine was created in 1977 under the aegis of the RRC. The difficulties the EWS confronted at the beginning subsequently persisted, given a shortage of funds, of transport and of properly trained staff. A proper EWS was crucial, however, if the RRC was to get beyond the hit-and-run character of its emergency relief efforts. Donor governments in the West could easily appreciate the EWS's importance, but many of them found the prospect of lending even indirect support to the Dergue singularly unattractive.

The vital EWS initiative was backed at the start by the USA (USAID), the United Kingdom (ODA) and the UN (WHO). The RRC plan, overall, was to identify not only regions in deficit, but those in surplus as well; and not only to monitor the country's food production as a whole (through EWS), but to bring relief to affected areas. The RRC would not just neutrally 'bring relief', it would also market grain and livestock, and promote transshipments of cargo from surplus to deficit areas. Most importantly, it would plan and execute the stockpiling of emergency food reserves: without stockpiling, good organization could achieve nothing, and drought would continue to mean famine. (It is hard to believe that this stockpiling began only in 1983, with the initiation of a modest FAO-backed 'Food Bank'.)

The RRC was much criticized as ineffectual. There was justification for criticism. For the members of the Dergue, personal survival was more important than relief and rehabilitation. The hardliners in the Dergue – firm on retaining territorial integrity – could count on majority support. Winston Churchill once claimed that he had not won power to abandon Empire. This was the rationale, the mood, of the Dergue, and of Colonel Mengistu. In such a mood, one may even sell one's country to save it. The debt to the Soviets for armaments received for the critical period 1977–9 was estimated to lie in the – for Ethiopia – unrepayable area of US$2b. The fiction of continuing repayment, however, has not been abandoned.

Relief and rehabilitation were important, but keeping the country together was regarded as more important still. To do anything meaningful in the way of relief, the RRC required external financial support. The first serious blow to relief operations fell in 1979, when (having shut off arms supplies two years earlier) the Carter administration withdrew support from the Early Warning System. British assistance had consisted only of two technical advisers for a term of two years. The upshot was that the vital EWS project within the RRC collapsed in 1980. A year later, UNICEF (the UN children's fund) rescued the EWS, and the programme was made operational from 1981. It is still far from perfect, but Ethiopia now has one of the best early warning systems against famine in Africa.

The discontinuation of aid from the West, especially from the USA, fits into a clear ideological pattern of opposition to regimes aided by the Soviets. Neighbouring Kenya, for example, has about half of Ethiopia's population. Kenya and Ethiopia remain on good terms and have almost certainly conducted joint military operations against the Somalis, despite the fact that Kenya provides base support for the US Rapid Deployment force in the Indian Ocean area – the object of which is to contain threats from states like Libya, Yemen . . . and Ethiopia. Kenya, in the period up to 1984, had one major crop failure; Ethiopia had at least four. But Kenya – the famine story came out in October 1984 – received three times the Western emergency aid promised to Ethiopia.

The question is often raised why a government such as that of the United States should believe itself obliged to assist African or other Third World states aligned with the Soviet Union. A part of the answer is that such 'alignment' is usually, perhaps necessarily, partial. Even at the height of the Soviet Union's transfer of arms to Ethiopia in 1977–8, Soviet and other Eastern bloc trade was only 2 per cent of the Ethiopian total. The West buys the bulk of everything that Ethiopia produces for export, especially coffee, which represented over 60 per cent of all Ethiopian export earnings in 1980, in 1982, and presumably in 1985.

Not surprisingly, most of the investment in Ethiopian cash crops – coffee, cotton, sugar, tea – has tended to come from Western and multilateral sources such as the EEC, the World Bank and the African Development Fund (ADF). The Soviets have invested in Ethiopia, for instance in petroleum exploration and gold-mining. But there is a severe limit to what they can do. The USSR is meant to be a global power, but the economic clout of the Soviets does not measure up to their politico-military status. With a population more than twice the size of Japan's, the USSR has a GNP which is only a touch ($2\frac{1}{2}$ per cent) larger. To make a different comparison, the population of the USSR is four and a half times larger than West Germany's, but the economy is less than half again as large. In sum, the Soviets have not the economic capacity to make the major contribution to Third World development which their political status implies, because their economy has only a fraction of the capacity of the West.

There is an inevitable Third World dependence upon the richer states of the West. But the dependence also runs the other way: the productive capacity, the employment, the surpluses and indeed Western economies as a whole would be severely diminished and disrupted were Third World outlets and resources suddenly blacked out. This involvement is well nigh irreversible. The African states are closely tied to their northern neighbours in Europe, just as Latin American economies are locked into those of North America. The economies of today's African states – even Ethiopia's – were originally conceived in a Western colonial embrace (often an act of rape), and all subsequent stages of African economic development have only reflected and reinforced this original penetration. It is understandable that many – not Africans alone – might wish to repudiate the economic ties inherited from the past 500 years. But the odds on doing this at all successfully appear fairly long.

Present-day African governments are sustained by a tax base which is grounded in established economic exchanges with the West. There is no obvious way of substituting some other tax base for the present one. To destroy African trade with the West, which is formally possible if practically out of the question, would almost certainly bring about the dissolution of those states most affected. A common way for a country to avoid political disintegration, where devoid of a tax base to maintain its territorial integrity, is to surrender territory to a foreign power for some strategic, military purpose of the latter, in exchange for receipts to keep the army, and thus the territory, in one piece. This is what

Somalia did, first with the Soviets after 1969 and then with the USA after 1977. This is a far from perfect answer and it usually leads states, thus militarily embraced, to seek a measure of relief and independence by renewing or extending links with the hitherto excluded super-power. This inclination for Third World states to break away from their patrons usually benefits the West, which can offer significantly more than military supply. The West is the most important source of investment capital and often the only possible market for Third World commodities.

The dependence of African economies upon exchanges with the West is not dissolved by embracing Moscow, whatever the reasons, good or bad. Similarly, Western responsibilities do not evaporate where African governments seek Soviet help. The Soviets, as we have seen, have relatively little help to give. Those African states with whom they have been closely associated, such as Sékou Touré's Guinea and Congo (Brazzaville), prove this. The most important assistance African states typically receive from the USSR is military. This has been true for independent states (like Nigeria in 1967–70, Ethiopia and Angola) attempting to contain secession and/or external attacks, as also for national independence movements (as in Mozambique, Zimbabwe, Namibia and South Africa) opposing colonial and/or racist rule. Western governments – most especially the USA – have been greatly unsettled by Soviet military assistance to national liberation and anti-racist movements in Africa, and often more unsettled still by Soviet military support to independent African governments seeking, themselves, to contain rebel movements. Since so much of this Western opposition is, on the face of it, irrational, it may well prove more affected than real. In all cases of Soviet military assistance to Africa, such assistance was sought from the East because it could not be secured in the West. When given by the East, aid has been justified on a combination of strategic and ideological grounds. The Soviets – whose aid has little developmental significance – for this reason have equally little staying power, of which all Western chancelleries are, or should be, perfectly well aware.

In the case of US aid to Ethiopia, the decision, by 1977, was to allow the country to sort itself out as best she might. Ethiopia needed military assistance, without which she would likely collapse. Even if the Soviets enabled Ethiopia to win the civil war, they would be unable to consolidate their hold in non-military, economic spheres. After all, Soviet relations with Sudan were reversed after 1971 by President Nimeiri. President Anwar Sadat expelled the Soviets from Egypt after 1972 (despite their vital backing for development). President Siad Barre was thinking along similar lines in early 1977; he finally expelled Soviet military personnel in November and cut diplomatic ties with Cuba. Even if the Soviets successfully intervened to save Ethiopia, they would necessarily further alienate Arab states in the process, render continued relations with Somalia exceedingly difficult, and still not find themselves securely ensconced in the country they would have made so great an effort to help. Such was the speculation that may reasonably be supposed to have passed through America's official mind in early 1977.

Given these circumstances, it is difficult to imagine that, at any point in the period 1975–8, the primary concern of the Dergue would or could have been with relief and rehabilitation. Attention was being paid to the problem, but the soldiers involved were paying more attention to apparently greater troubles. The Relief and Rehabilitation Commission, when created, had very limited resources. It was not at first run as well as it might have been had better administrators and more resources been available, but the evidence does not support the conclusion that it was misappropriating relief assistance or over-reporting Ethiopia's needs, whether to generate a surplus for the army or for other purposes.

A number of charges were made against the RRC by various individuals, most importantly perhaps by Lord Avebury in Britain's House of Lords on 7 June 1982, to the effect that food aid from the EEC and UNICEF had not reached the poor for whom it was intended, but was used by the Ethiopian military instead to provision their front-line troops; and that it was sold by the Ethiopians to the Soviets to pay off their military debts. The only hard evidence for these charges was that Ethiopian army units, overrun by either Tigrayan or Eritrean guerrillas, were found in possession of flour, oil and other food in containers marked to show that these were gifts of the EEC or of the World Food Program, or of the International Committee of the Red Cross. Most of these charges originated with the guerrilla movements themselves.

The most inaccessible areas which the RRC was charged to reach were the two northern provinces of Eritrea and Tigray, both beset by civil wars since 1962 and 1975 respectively. Relief workers on both the rebel and government sides were killed by opposing combatants. On the Ethiopian side, food was conveyed into Tigray and Eritrea and into neighbouring Wollo by military vehicles and under military escort. The rebel movements sought to wrest food supplies from the government.

The government restricted supply, in terms of the amount they transported and stored in insecure areas. Thus, supplies into rebel areas were more limited, and were always associated with some Ethiopian military presence. This does not, however, establish that there was systematic misappropriation. Some overseas food aid did turn up in Somalia and in Sudan. This sort of thing occurs commonly where the type of food supplied is not well suited to the target population. The result is that foreign high-protein wheat, for example, is exchanged for local and more suitable cereals, thereby bringing such wheat onto the market. The EEC regards this procedure as normal – indeed, as advantageous.

On 27 March 1983, an article by Simon Winchester in London's *Sunday Times* bore the headline: 'Food for Starving Babies Sent to Russia for Arms'. His authority for this conclusion was a nameless Ethiopian official who was said to have defected. EEC grain, Winchester claimed, was relabelled at the port of Assab on the Red Sea and then reloaded onto homeward-bound Soviet ships. These claims were, of course, scrupulously checked by the donors – the UN, the governments concerned, non-governmental organizations and the EEC. On 14 April 1983, Edgar Pisani, the EEC's Third World aid Commissioner, announced in Strasbourg before the European parliament that EEC food for Ethiopia had been misdirected neither to the Ethiopian soldiery nor into the bread bins of the USSR. EEC investigations established that the Red Sea ports are small, that loading and reloading cargo is slow and conspicuous, that these ports are not subject to military control, that numerous Westerners work in and visit them, that eighteen different cargoes had been checked in the course of two years, that the 750 storage points throughout Ethiopia (to which food was trucked) were subjected to random checks, and that evidence overall of peculation, in the ports and up-country, was nil.

The basic problem in northern Ethiopia is less to do with peculation than with how to provision drought-affected areas which double as theatres of war. The Dergue employs the RRC to attend to the starving by setting up distribution points along major roads. These points have to be defended by the military against an enemy not readily recognized as such, nor differentiated from civilians. The rebel areas of Eritrea and Tigray, to the extent that they supply themselves, must depend upon access to Sudan. Very little can be got to the northern population in this way because of inadequate infrastructure and aerial interdiction by Addis Ababa. The Dergue controls the towns, not the countryside – in

Eritrea, in Tigray, even in much of Wollo; hence the Dergue cannot directly reach drought victims except by attracting them to the distribution centres subject to its control.

The war between the Dergue and its various rebel opponents is not simply a war between classes or rival parties. It is also a war between regions, and thus between regionally entrenched ethnic groups. In wars such as this even the little underlying sympathy for destitute civilians that one might expect to find among an ill-trained conscript army is, as often as not, absent. An able-bodied young man who turns up in a northern town like Makele, speaking Tigrinya and seeking RRC assistance, begs by his very presence to be considered 'one of them' – a rebel. He may have a lot of explaining to do. He may be roughed up. He could be conscripted. He could be taken and sent to Western Ethiopia. It could easily be worse. Healthy young women who enter upon such a scene are abused, with little prospect of appeal to higher authority. To observers who protest, it will be officially pointed out, often more with resignation than animosity, that 'there is a war on'. Such men and women, unless they really have no choice, will not come into the Dergue camps. They are afraid and would prefer to take their chances in Sudan, even if it means walking all the way. But a lot of people do come – the women, young with child or ageing with family, premature crones; the children, the old men, the sick and disabled. The Dergue administration is best placed to help, but the Dergue wants to win a war.

Given that the RRC is too vulnerable to operate on a small-scale basis throughout the countryside of the northern provinces, and given that the rebel movements are unable to ensure that the people they profess to control in these regions are properly provisioned, common sense and humanity would dictate agreement upon some form of ceasefire between the two sides in order to halt the unnecessary loss of life. The Dergue is unhappy to offer a ceasefire for as long as the rebels insist either upon secession or self-determination (which in Tigray might or might not lead to a vote for independence as opposed to some loose form of federation). The rebels would welcome a ceasefire so long as it does not require them to surrender their autonomy and in so far as it strengthens their hand. The Dergue will not yield on the principle of Ethiopian sovereignty. The rebels will not yield on the question of their own autonomy of decision. Both sides are effectively responsible for the deaths of thousands because they regard such deaths as preferable (in the one case) to the loss of Ethiopian territorial integrity or preferable (in the other case) to being joined in some sort of union with Ethiopia.

There is no question but that Ethiopians are dying in the north because combatants on both sides place a higher value upon future political structures than upon immediate human suffering. The rebels are in a weaker position and thus more readily make *ad hominem* charges against the Dergue in order to undermine the latter's claims to legitimacy. The real problem is not that either side is violent or engaged in an orgy of pilfering or misappropriation, but that their radically different ideas about the future structure of the region are directly exacerbated by the ingrained antipathies between the great powers, who should be seeking to mediate these differences, not to exploit them. The people who are being killed by these rivalries are not Soviets, or Americans, but the peoples of the Horn. The boys in the ranks of the Ethiopian army, members of the Dergue itself, militants of the various rebel movements are dying in order to demonstrate which of the super-powers is blessed with the more powerful musculature. No one pursuing the Ethiopian policies of the Carter administration, or that of Mr Reagan, or that of Mrs Thatcher, is entitled to call Colonel Mengistu 'callous'.

A superficially more plausible complaint than that the Ethiopians have used food aid to support their war effort is that the RRC over-reported Ethiopia's emergency food needs. Here, too, however, analysis of the evidence shows no clear support for the charge. The limited resources of the RRC (especially the EWS) make really accurate assessment of conditions difficult to supply. But even when this con-sideration is taken into account, what emerges is that the RRC displayed a marked inclination to under-report rather than the reverse. If ever there was over-reporting, this could not have occurred before 1985.

By early 1983, the famine in Ethiopia was already at a very advanced stage. On 4 April, an important meeting was held in the RRC conference room in Addis Ababa. The then Chief Commissioner of the RRC, Ato Shemelis Adugna, claimed that the Ethiopians threatened by drought and famine numbered 4,008,000. He stated this on the assumption that the population was only in the order of thirty-two million people. But within eighteen months, following the first proper census ever conducted in Ethiopia, it was established that the population was over forty-two million people. Mr Shemelis' figures would have been seriously out of kilter for this reason alone. In any event, his figures plausibly showed drought victims to be roughly three times more numerous in the north

(Wollo, 850,000; Tigray, 1m.; Eritrea, 701,000; Gondar, 489,000 – a total of 3,040,000) than in the rest of the country (a further 968,000 victims). But the Chief Commissioner's numbers appear strikingly optimistic in retrospect. Mr Shemelis, taken to be well-meaning and honest if not overwhelmingly efficient, later left his position. But the optimistic under-reporting of drought victims by the RRC continued none the less. It represented, clearly, a more systematic phenomenon than can be accounted for by reference to the alleged foibles of some particular administrator.

In November 1983, at a time when the number of drought victims was increasing, the RRC said that they had dropped from Mr Shemelis' four million plus (given over half a year earlier), by 208,390, down to 3,799,610. The RRC November report also said that the numbers of drought victims in 1984 would drop an additional 200,000-plus, to the level of 3,559,700. As it happened, the numbers affected by drought in 1984 came closer to six million than to three and a half million people. The RRC did not over-report, as charged in the Western press. Until well into 1984, it under-reported – in the case cited, by as much, perhaps, as 40 per cent. This could just have been a mistake. If so, it was not that of just one Chief Commissioner. The RRC projections, in fact, are systematic enough to lend colour to the hypothesis that they were generated more by policy than by miscalculation.

Doubtless, there was no proper appreciation of the real extent of the tragedy unfolding in Ethiopia. The foreign press, governments and aid agencies were invited to believe that the crisis was of manageable proportions, and could, with modest assistance from outside, be contained. The revolution, after all, was meant to embody a triumph, not a tragedy. The 1983 figures given by the RRC conjure up the image – wholly fictional – of a steady attack upon, and reduction of, the misery overwhelming the country. The tenth anniversary celebration of the overthrow of Haile Selassie would fall in October 1984. These celebrations were acutely political – designed to affirm Ethiopia's endurance, to inspire the people, indeed, to spit in the eye of the subversive and secessionist enemy. Colonel Mengistu, after all, was still waging war. Every time one enemy was swatted down, up sprang another.

In 1984, the new Chief Commissioner of the RRC, Dawit Wolde Giorgis, had a problem on his hands. He was a young man – quick, competent and firm. His job was to handle foreign donors, and to make them believe him. But he would also have to compel greater attention within the Dergue – which meant from Mengistu, more war leader than a dispenser of charity. By the beginning of 1984, the RRC was aware that the tragedy confronting Ethiopia was far more portentous than had been suggested by any previously published figures. To compel recognition of this within the Dergue was also to seek an adjustment of priorities at a difficult time – from 'Ethiopia First' to succouring the needy. The appointment of Dawit was the first sign of a notching upwards of the humanitarian priority. At the beginning of October 1984, the tenth anniversary celebrations behind him, Dawit made another urgent appeal to donor agencies for emergency relief assistance. His real problem, in fact, was to get Mengistu on his side – and keep him there. The extraordinary and embarrassing publicity generated by the Buerk and Amin BBC TV coverage of starving Ethiopia was of crucial importance in forcing Colonel Mengistu – in the same way as it forced other governments, such as Mr Reagan's and (less so) Mrs Thatcher's to reshape their priorities. By the end of October, Mengistu had been persuaded to assure emergency relief workers that every effort would be made to assist them in their tasks.

It may well be that governments are far less feeling than ordinary citizens, in whatever worlds we find them, First, Second or Third. Certainly there is little evidence that developed states have seriously taken to heart the fundamental task of assisting with development. It is in the direction of development that a crisis like Ethiopia's must point. It seems improbable that any new directions will be taken unless a groundswell of popular support in developed states demands it. Without the push for development, all of Africa, with its teeming life, human and other, strung out over wildly varied habitats, with its extraordinary agricultural and mineral potential lying upon the land and thrusting into its bowels, with its colour and vivacity and warmth, is set to become a vast refugee camp, filled to the brim with impoverished and restive people, whose seismic desperation, unrelieved, may well shake the foundations of politics in our century.

# 5 Colonialism and Catastrophe: Ethiopia, Sudan and Mauritania

The war launched by the Eritreans in 1962 against the central government in Ethiopia is now stretching into its third decade. The other domestic conflict in Africa with which Ethiopia's is most commonly compared is the Nigerian civil war of 1967–70. The 'Biafran' war claimed a million lives. It involved a significant attempt by Nigeria's Ibo-dominated south-east to secede, just as the Hausa-dominated north threatened to do in 1953, and the Yoruba-dominated west also threatened in 1954, and the north again in July 1966. Nigeria is three-quarters the size of Ethiopia. Hundreds of languages are spoken within its borders. At the time of the Nigerian war (as in the Ethiopian case) the West would not supply the centre with the military hardware required to hold the state together.

Unable to secure Western support, in 1967 Nigeria turned to the Soviets, and with vast oil revenues (which moved from 2 per cent of the total value of exports in 1963 to 66 per cent by 1972 to 80 per cent in 1985) was able to pay for the fighter planes, 122mm. guns, ammunition, etc. which she sought. This war cost Nigeria something short of $1b. Biafra was supplied by Ian Smith's Rhodesia, by South Africa and by France, while Portugal partly bankrolled and facilitated gun-running into Biafra. The United States did not allow itself to become directly involved (although the USA and especially Britain were significantly involved in diplomatic efforts to keep the country in one piece). The Nigerians were sufficiently well-heeled not to require to offer the Soviets strategic military emplacement in exchange for the war *matériel* obtained. It was fortunate, too, for the Nigerians that they had no neighbours with anything like their wealth, population and power; and that support for the secessionists was not supplied by any of Nigeria's mainland neighbours. Ending the Nigerian war was painful, but it was manageable.

Ethiopia's position is far more complex. The multiple conflicts in which she has become embroiled appear interminable. The Ethiopian highlands endure an almost automatic notoriety in that the Blue Nile has its source in Lake Tana: northern Sudan and indeed Egypt would be all desert *grosso modo* if these waters were diverted. Ethiopia, understandably if unfortunately, has virtually never had good relations with these northern neighbours. It is equally unfortunate that Ethiopia's position is of such vital importance to the super-powers: their compelling interest in the free movement of their shipping through the Bab El Mandeb choke point, which is dominated to the west by Ethiopia, makes these powers inordinately sensitive to Ethiopian domestic developments. Ethiopia's population and army are the largest in the region. It is a situation that renders Ethiopia a potent ally but, by the same token, a daunting adversary. None the less, Ethiopia does not dominate Eastern Africa in the way that Nigeria dominates Western Africa.

Ethiopia is almost twice as large in area as Somalia and seven times more populous. Although, before the 1983 famine, she enjoyed (on UN figures) generally better health and education, even then her GNP per capita was about half that of Somalia.

In comparison to Sudan, Ethiopia is about twice as populous, but half as large; she is consistently less well placed in education and health and her per capita GNP is only a little more than a third as high as Sudan's.

In comparison to Kenya, Ethiopia is about twice as large, both in area and population, but otherwise suffers from the comparison, with a per capita GNP one-third of Kenya's, life expectancy at least ten years less, and levels of literacy and health care much inferior.

By any standards, then, Ethiopia is poor, and this applies even within the Eastern African context. In such a country, the government is more likely to have to resort to force to hold itself together, thus multiplying the burdens imposed upon already burdened domestic populations.

Ethiopia is intrinsically more friable than Nigeria. She is geographically larger, her terrain is more broken, she has a lower population density and greater phenotypic variety, to be set off against a poorer infrastructure, roads few and bad, exiguous communications, with historically grounded and deeply rooted conflicts, especially between Christian and Muslim, to which may be added significant exposure to competing colonial challenges (from Egypt, Italy, Britain and France). Ethiopia, although bled by colonial states and even defeated (by a British army of 32,000 men under Napier in 1868 and by the Italian army under Badoglio in 1936), never remained captive long enough to have a modern infrastructure imposed upon her or to derive a larger, positive identity from an alien presence.

The Ethiopians struggled long and well to maintain independence against colonial powers. But while they did better than most other African polities, domestically they were but an anachronistic feudal autocracy. Ethiopia was technologically capable of sustaining a limited ruling elite only on the basis of semi-subsistence agriculture. Polities of this kind are generally not well integrated, nor can they be. They can serve as no match to industrial polities with their concentrated and homogenized populations, with the massive productive capacities attending them.

Eastern Africa, and most especially the Horn (principally Somalia, Djibouti and Ethiopia), is by far the most complex of Africa's regions. In the Horn, local sensitivities count for much. But local conflicts have always been exacerbated by the competitive interplay of external powers seeking strategic advantage in the region. Although this external presence cannot be wished away, it would be naïve to suppose that it has brought much good – whether today or a century ago – to local people. The great powers must be mutually opposed. And they will hotly recruit clients in sensitive areas, clients who are already mutually opposed. In the last century, competition for control of the Horn was essentially between Britain and France. In this century it is essentially between the USA and the USSR. The more evenly matched these powers, the more devastating their combined effect.

The regional tensions which envelop Ethiopia are no less daunting. Sudan has inevitably been concerned with developments across her common border with Ethiopia. Refugees have passed from northern Ethiopia into Sudan and from southern Sudan into Ethiopia, and continue to do so. In 1984, according to the head of Ethiopia's Rehabilitation Commission (RRC), as many as 100,000 could have come from Sudan into Ethiopia. In the 1960s and subsequently, Sudan provided support to rebels in Eritrea, just as Ethiopia supported the rebels in Southern Sudan (Upper Nile and Equatoria). Somalia, in the 1960s and after, provided support to ethnic Somali rebels in the Ethiopian Ogaden. She directly invaded Ogaden in 1977. Ethiopia is joined with Kenya in a common defence treaty against Somalia. She also provides covert assistance to anti-Mogadishu rebels. Leaving Sudan and Somalia to one side, many Arab states across the Red Sea, like North Yemen (YAR), Saudi Arabia, Syria, Iran, Iraq and Kuwait, have at various times provided significant support to anti-Ethiopian rebel groups in the Horn, and much of this continues.

What dramatically complicates domestic and regional tensions is the effect of super-power rivalry. The Horn abuts upon that region which contains the world's most extensive oil reserves. Sudan and Ethiopia themselves have or are likely to have significant reserves. The West depends upon Middle Eastern supplies; Soviet dependence upon supply of Persian Gulf and North African oil has picked up. It is largely via the sliver of water comprising the Red Sea and the Suez Canal that the trade of the Atlantic and the Mediterranean is linked to that of the Indian Ocean, the Pacific and the West coast of America: fully three-quarters of Western Europe's raw materials travel via Bab El Mandeb and the Isthmus of Suez.

Similarly, if less crucially, the Indian Ocean route is one of three means of linking the east of the USSR with its west. The northern sea route is passable in only three months of the year. The trans-Siberian rail route is immensely long (the USSR being about 5,000 miles across), is subject to the vagaries of singularly harsh winters, and would be acutely vulnerable to external (Chinese) attack in case of war. Apart from questions of trade routes and choke points, the USA has nuclearized the Indian Ocean, and thus threatens the Soviet Union in a weak spot. The only way to counter such threats is by attempting to move into the regions whence these threats originate. Hence the Soviet concern, at various times and with variable success, to cultivate Egypt, Sudan, the two Yemens, Ethiopia, Somalia, Uganda and Kenya.

The concern of the USA, as stated before Congress by the American Secretary of Defense, Casper Weinberger (1983), would serve almost equally well (suitably amended) as a statement of Soviet policy: 'to prevent the spread of Soviet influence and the consequent loss of freedom and influence which it entails; and to protect Western access to the energy resources of the area, and to maintain the security of key sea lanes to this region'.

Britain, stung by Napoleon's 1798 lunge for Suez (with its attendant threat to continued British suzerainty in India and the East, imperilling British lines of supply and communications round the Cape), fitfully set about the business of bringing the Red Sea littoral and environs under her control. Egypt was taken, then Sudan; Ethiopia was broken; Eritrea allowed to a weak European protégé; northern Somaliland invested; a protectorate established in Uganda; Kenya colonized. On the other side of Bab El Mandeb, a similar process was unfolding: in 1802 a treaty with the local sultan, which left commercial traffic insecure, followed by military occupation of Aden in 1839; the gradual extension of control along the eastern shores of South Arabia, to Oman and into the Persian Gulf; steady opposition to Turkish pretensions in Yemen and Arabia; the orchestration of Arab revolt against the Turkish ally of Germany in the First World War; support for Ibn Saud in the construction of Saudi Arabia; the absorption of Jordan into the British zone of influence following the Sykes–Picot Treaty of 1916; and the 1920 absorption of Palestine (hived off from Syria) as a mandated territory.

The logic of global dominance dictated creeping British control of Suez, of the Red Sea, the Straits of Bab El Mandeb, the Eastern African littoral down to Tanganyika, and the South Arabian littoral to the Persian Gulf. The Second World War effectively signalled the end of this empire, culminating in British retreat from 'east of Suez' by 1968. The American eagle, at the close of the Second World War, replaced the British lion as the arbiter of man's fate in Europe. This role was gradually extended into Egypt, Arabia and Eastern Africa. The challenge to British dominance was not from the USA: the switch here was in the form of a relay race. The challenge to Euro-American dominance was from the Soviets. The French made their stab under Napoleon. The Germans took the plunge under Hitler and thereafter were split asunder. British emplacement at the Straits of Bab El Mandeb was challenged by Nasser, supported by the Soviets, and these pressures culminated in Britain's 1968 withdrawal from Aden. Insufficient Anglo-American support for Somalia's irredentist claims invited Soviet entry after independence (1960). Soviet influence in Somalia reached its apogee in 1976, being reversed by the 'switchover' of 1977. Soviet influence was expunged from Somalia by virtue of the critical military assistance the USSR accorded to Ethiopia (under attack from Somalia). US influence in Ethiopia, by virtue of the refusal of military assistance, was equally expunged. The USA, by virtue of the physics of super-power relations, was suctioned, in its turn, into Somalia.

The world's two major powers, the USA and the USSR, like their predecessors in the Mediterranean, Britain and France, have chosen to square off (not least importantly) in the Red Sea and Indian Ocean regions. The USA elaborated what was at first dubbed a Rapid Deployment Force to shore up allies in places like Egypt, Sudan, Israel, Arabia, the Persian Gulf and Kenya. The USSR managed to secure strategic emplacement in Ethiopia and across the water in Aden.

The struggle for supremacy in the region of the Horn continues unabated. Soviet and American policies are run along parallel lines. The US attempts to maintain its dominance, the USSR to break that dominance. Neither side wants war among its clients. The USA, to please its Arab allies, would be pleased to break Ethiopia, juggle its components, exclude the Soviets from the region and construct some form of local peace on this basis. The major difficulty here is that America's Kenyan ally is opposed, perhaps terrified, by the prospect of such a development. In the long term there is little that Kenya can do to resist American policy in this regard, and there are small signs that Kenyan officialdom may have begun to resign itself to the business of backing away from Ethiopia. As for the Soviets, in 1977 they tried their hand at controlling the region by attempting to persuade their clients of the time, Ethiopia and Somalia, to break bread together as fellow socialist states, to avert war, even to form a federation with Yemen (the PDRY).

Super-power rivalry, then, continues. The powers can no longer fight among themselves, so they interpose their neo-colonial hoplites, who in turn test the new, sub-nuclear weapons and strategies. Where the Americans despatch their forces, as to Vietnam and Lebanon, the Soviets keep clear. Where the Soviets deploy their forces, as in Afghanistan and Ethiopia, the Americans keep clear. The two sides avoid discussion because both wish to avoid concessions. In consequence, the people of the Horn suffer. They suffer because of their own deep-seated animosities. And they suffer because the great powers, in their wisdom, find it politic to exacerbate these animosities, to nurse and to arm them. Drought strikes recurrently, but the limited energies of Ethiopians and Somalis and Sudanese are harnessed to the machine of war. Refugees, desert, disease, all steadily encroach. Ample rain returned to many areas in late 1985, with harvests towards the end of the year. But of course the rain, sooner or later, again will fail. And should the failure of political will continue, then famine may be expected to recur on an even grander scale.

In Ethiopia, over the period 1958–78, out of one hundred districts, only seven escaped famine. During this time, one researcher estimated, more than five million Ethiopians died. Reviewing the history of these developments, it becomes clear that the scale of each successive famine swells. There are remissions, but with each new onset the effects become the more devastating. At the height of the latest famine, in October of 1984, the toll of dead was probably about 7,000 fatalities daily. By 1985, every region in Ethiopia was affected. The failure of the 1982 harvests was the last of a series of events triggering shock waves that reverberated, in the end, not only throughout Ethiopia but into neighbouring Sudan, Djibouti and Somalia.

Sudan shares with Ethiopia 1,366 miles (2,200km.) of only partially demarcated border, running roughly north to south. To move westward from Ethiopia to Sudan is to drop from 9,000 feet or more to 3,000 feet or less. It is a difficult border to cross. Hundreds of streams – the most famous being the Rivers Atbara and the Blue Nile – descend to bring life to the lowland peoples. But as with water, so with people: movement is mostly one-way. Drought, war and famine, which forced so many Ethiopians from their homes and villages, also forced hundreds of thousands of them into neighbouring states. The outflow of Ethiopians into Sudan was estimated in April 1985 by the US government at between two and four thousand daily.

In March 1985, there were already 66,000 Eritrean refugees in the Wad Sherife camp, 18km. into Sudan, near Kassala. Within months, this figure exceeded 140,000. Thousands of Eritreans were trucked south to Girba to get them well away from the Ethiopian border. In late 1985 there were 500,000 Eritrean refugees in Sudan. Tigrayans, too, poured into the country. By March, 82,000 of them were being attended to in a camp south of Wad Kowli. By mid-year their numbers exceeded 100,000. As many as 12,000 Ethiopian Jews had, along with the rest, tumbled pell-mell into Sudan. In Ethiopia they are styled 'Falashas', which in Amharic signifies 'stranger'. But since they have been about since well before Addis Ababa was built or Columbus sailed into the Caribbean, they understandably view the term as ridiculous or offensive. The Israelis scored a great propaganda victory (with the help of the USA and of the Nimeiri government) by airlifting them out to Israel between November 1984 and January 1985. All told, by April 1985, there were 750,000 Eritrean and Tigrayan refugees in Sudanese camps, the numbers constantly growing. Others came from Wollo. They travelled on foot, sometimes by truck, usually but not always at night, on a journey that could take as long as two months.

In the second quarter of 1985, even as the effects of the famine in

Ethiopia became more chronic, the actual death toll declined. With the decline in mortality, not only was there a decline in world attention, but it also switched – in the first instance to Sudan. The seven African states worst affected by drought and its consequences in 1984–5 were Ethiopia, Sudan, Angola, Chad, Mali, Mozambique and Niger. In December 1984, to cope with the spreading disaster, the UN Secretary-General, Javier Perez de Cuellar, set up a UN Office for Emergency Operations in Africa (OEOA) appointing Bradford Morse – chief of the UN Development Program (UNDP) and former US Republican Congressman (Mass.) – to head it. Maurice Strong, a Canadian multimillionaire and former head of the UN Environment Program (UNEP is based in Nairobi), was appointed as executive (or field) coordinator of UN emergency operations in Africa. Kurt Jansson was required to mind the store in Ethiopia (as UN Assistant Secretary-General in charge of OEOA there), while Strong had to keep an eye cocked on developments in the continent at large.

Sudan was seen to be slipping into the same slough of despond as her neighbour. A collapsing economy, strikes and food riots presented the final challenge to a tired and fossilized government desperately struggling to save itself from internal dissidence and international economic pressure. While the Nimeiri government in Sudan strove to rescue itself, it was not well placed to attend to the rescue of its citizenry, among whom as many as a million children were under threat of death by starvation and related forms of extirpation. On 15 March 1985, from Geneva, Maurice Strong summarized the position: 'In Sudan, the lives of a million children, one in every six, may be lost this year.' On 6 April 1985, the Nimeiri government was toppled by the upper echelons of the military, in order to steal a march on restive elements among the lower ranks.

In Sudan, as in Ethiopia, drought, war and debt hammered local populations. Hundreds of thousands were flung across the borders, not only from Ethiopia, but from Chad and from Uganda. The total number of foreign refugees in Sudan in July 1985 was 1·3m. There was a significant inflow of refugees from Chad into Sudan's western Darfur province over 1984, and early 1985. Their numbers reached 121,000. The Aserni camp, in Sudan's far west, was set up in January 1985 to hold 10,000 refugees, and had received, within two months of opening, two and a half times this number. These were mostly nomadic or semi-nomadic people whose traditional water supplies had disappeared, whose camels and goats had died, who were forced over the border by civil strife and who could expect no help from a government with holes in its pockets and a civil war to feed. The peoples in the west lay about 1,200 miles by air from the relief food available at Sudan's Red Sea anchorage at Port Sudan, and little of that food got through. It is to be presumed that the vultures at least were well pleased with conditions: observers claimed never to have seen so many birds of prey, darkening the sky to the horizon. There are no statistics to be relied upon. Into the 1985 autumnal gloom of the north, sunlit relief workers in the Sudan transmitted figures reaching 150,000 dead for these distant, western marches. But by October 1985, the rains had come, churning desert into mud, making communication (certainly transport), and so relief, all but impossible.

The 1985 drought affecting the Sudanese provinces of Darfur (in the west), Kordofan and Red Sea had lasted for four years. Harvests were poor. Food was scarce. Crops and livestock had been decimated. As many as 4·5m. people were affected. Ethiopians were pouring into the south-east of Sudan, Chadians into the north-west, Ugandans into the south – by January 1985 there were as many as 195,000 Ugandan refugees and 5,000 Zairians. The peoples across these borders – Ethiopia and Sudan, Uganda and Sudan, Chad and Sudan – entertain a lively perception of their common ties. Not only were 'foreign' refugees on the move. Sudanese themselves were joining in, and in the border areas, as often as not, sought to pass themselves off as non-Sudanese, in order to enjoy the dubious benefits of refugee status.

The north of Sudan is mostly desert. Water resources were shrinking. Nomads fell pell-mell upon the rivers. Women, children and old men collected in the camps and towns along the way. Younger men drifted off to save themselves or to rescue their families or to perish. Animals disappeared, the price of grain skyrocketed. American grain reached some, by means of three helicopters, costly to operate, covering a vast region. People ate the leaves of the fever trees and boiled the poisonous berries of the mokheit bush, but many survived against all the odds.

Within a fortnight of the overthrow of General Nimeiri, the new Sudanese leader, General Sewar El Dahab, announced from Khartoum (20 April 1985) that 1·5m. rural Sudanese had migrated to towns in search of food. By July 1985, food reserves from the poor 1984 harvest were exhausted. Additional foreign food aid was now urgently, officially and publicly sought. Sudan had virtually no food reserves to fall back on. She bore an external debt of over US $11b. The interest rates being paid to Western banks for loans meant to fund development were inappropriately high. Half of the population, 11·4m. people, were suddenly looking up the barrel of starvation. As many as two million Sudanese were already working abroad, mostly across the Red Sea, in oil-rich Muslim states like Kuwait and Saudi Arabia, as teachers, technicians and the like: there were few jobs at home, and even those abroad were contracting.

The same forces then that inexorably pulled Ethiopians west and Chadians east into Sudan, also magnetized the Sudanese themselves, and in the arid north-west (in Darfur and Kordofan) folk were drawn basically further east and south. The failure of the 1984 rains was but one in a series of failures that stretched right across the country, from the north-west (Darfur) to the north-east (Red Sea province). The crops that failed, the beasts that died, catapulted a multitude of people towards railways, the major roads and most importantly onto the banks of the Nile and into the irrigated Gezira region below Khartoum.

In Darfur, with an estimated population of 3·1m. people, among them about half a million nomads, the 1984–5 drought had begun in fact as much as twelve years earlier, becoming especially severe from 1980. The Sahara desert (including Sudan's north) is advancing at a rate of between four and six miles yearly. It is gradually absorbing Sudan's northern farming and grazing lands. The people most severely affected, 13° north of the Equator, are gradually being drawn (on the whole) south and east of this line. Their displacement increased population densities in the more fortunate areas, and intensified demand for and pressure on remaining pasture and water.

In June 1985, more than 80 per cent of Darfur's population stood in need of food. By August, the UN estimated that only 10 per cent were being reached. Road and rail links to western Sudan, including Kordofan, were frail. The rail system would be required to transport nine times the tonnage that, on the best estimates, it could cope with. USAID depended on the private US trucking agency, Ark El Talab, to deliver the goods, but private enterprise in this case had been slow off the mark. The EEC 'airlift' was hobbled by the low carrying capacity of planes and by fuel problems. Everyone feared that the transport system would collapse with the demands being placed upon it and particularly with the onset of the rains. So it proved. A major bridge disintegrated at Nyala, on the line to northern Darfur, in the first week of July 1985. Two locomotives were pitched into a ravine with more than half-a-dozen goods cars. Another stretch of rail, under the burden of a train loaded with vital fuel for Darfur, simply dissolved into the clay on which the sleepers were laid: Darfur was not to be salvaged.

The tragedy here was not of the starkly visible sort associated with the compact misery camps of Ethiopia. The statistics on Ethiopia were unreliable enough; those on Sudan were worse. Yet the gross scale of events was not to be denied. Darfur, north and south – with a total area of about 200,000 square miles (over twice the size of the United Kingdom) and a population density of only about fifteen persons per square mile – was a land of scrub and desert, dotted with villages, served by no roads for the most part, and thus not readily mapped. The same applied to neighbouring Kordofan, a quarter nomadic, similar in geographical extent and population density. The pattern we know to expect was that which held. There was considerable emigration. Water for some villages (as near the town of El Fasher in northern Darfur) was as much as fifteen miles distant. People were forced back upon berries, roots, seeds: 'famine food'. Malnutrition spread, particularly among the childen. (It is estimated that, in the period February–July 1985, out of a total population of 1·6m., as many as 8,640 children died in northern Kordofan alone.) The price of animals (difficult to maintain) dropped. The cost of grain rose. Migration among males accelerated. The old, the young, the women, stayed behind.

As usual, the people who were hardest hit were the nomadic pastoralists. The nomads produce meat which they sell to villagers in exchange for grain. Being mobile, they are better at adjusting to forced migration than are agriculturalists, and can more easily make their way to distant distribution points. But farmers at least receive higher prices in famine conditions for such grain as they sell. Pastoralists, by contrast, always lose. Civil servants, on fixed incomes, suffer terribly by the increased cost of grain, but at least meat (for a time) is cheaper, and those on salaries, being in urban areas, have easier access to the emergency supplies brought to these centres. But pastoralists, once they have sold their animals, have neither these to feed on, nor land to work, nor jobs to steady them. They, more than the rest, were (and are) forced to the wall.

But not only were nomadic pastoralists in strife. Many migrant labourers from the south – Shilluk, Dinka, Nuer – had moved into southern Kordofan, following the traditional pattern, in search of work, at a time when less was to be found. In addition, there were more than usual, since many were escaping the renewal of civil war in the south. Sudan was receiving war-cum-drought refugees from both Ethiopia (to the east) and Chad (to the west) but also from her own mismanaged southern provinces further up the Nile.

Darfur, the worst affected of Sudan's regions, was not always as desperate as in mid-1985. Fifteen years before, it was able to export food (peanuts, potatoes, goats) to neighbouring states, as well as to neighbouring provinces. Food was taken from the region. But virtually no development money was put back in. Then the cost of oil exploded in the 1970s, inhibiting export initiatives. Insecurity in nearby states (Chad, Central African Republic) did not help. Insecurity within Sudan itself helped least of all. The country got bogged down in a civil war virtually from independence up to 1969. The leader (Nimeiri) who stopped it then lost his way much later (May 1983), and opened the second chapter in Sudan's civil war between north and south. The war drew off scarce resources that might otherwise have been used for development in areas like Darfur and Kordofan. Sudan's foreign debt, independently of the war, grew to enormous proportions, further siphoning off any prospect of substantial developmental assistance for all regions save (basically) the centre. Here, precious foreign exchange was borrowed and expended on expensive fertilizer and machinery for massive irrigation projects to develop cotton, especially, but also sugar. After 1974, interest rates went through the roof. Manageable loans for continued development were no longer feasible. Sudan's initially successful development of cash crops for export correspondingly declined.

During the severe drought of 1984-5, the Nile, depleted as it was, attracted large numbers of migrants. The Northern Region of Sudan (Northern and Nile Provinces) is mostly desert anyway, so local populations were already settled along the river. Irrigation is standard, but the low run-off from the Ethiopian plateau far to the south meant depleted water supplies, which, together with scarcity of fuel to work irrigation pumps, reduced crop yields. Northern agriculturalists were not so vitally affected, but pastoralists were. Large numbers (20,000) of the nomadic Kababish, for example, alien to the riverine folk, and badly hit by the drought, were drawn into the Debba area.

Further south, Sudan's most fertile region, Central, always attracts considerable migrant labour. Central Region contains the Gezira and Rahad irrigation schemes and, because it is not a drought zone, it reflects conditions in the rest of the country. After August 1984, there was a sharp jump in migrant inflow, estimated at about 300,000 people, some markedly destitute, many intending to settle for good. People were coming in from the devastation west of the Nile. As early as March 1983, villages like Abu Ferawa, Abu Roubka, El Gheseta and Tindelta, south of Kosti (in White Nile province) and west of the Nile (as little as thirty to fifty miles away), seem to have been decimated. Village populations were reduced by factors of 12 to 18, with no more than 200 folk remaining. The wells had dried up, children were dying, adults were sickly, animals wiped out. By October 1984, many who were down to their last were trying to make their way across the White Nile, in part to the dozen camps that existed at this time in the region. The capital, Khartoum, and its twin city, Omdurman, were not spared: orphans and vagrants were on the increase and many of these people (like the rural Ethiopians who descended on Addis Ababa) were simply trucked back to where they came from. In Khartoum people were required to form lengthy queues to secure their daily ration of *kisra*, the local sorghum-based bread. In the eastern provinces, the situation, tied into the Ethiopian débâcle, was appalling.

It is a common complaint that problems cannot be resolved by 'throwing' money at them. What is clear is that nothing is resolved where resources are not intelligently allocated. Sudan, and indeed all of the Sahel, stretching from the Atlantic to the Red Sea, has been relentlessly assailed by drought for year upon year. Dune-fixation, shelter-belts, the development of appropriate plants (for food, income or fuel) have been needed for years, and have been seen to be needed. But less than 2 per cent of foreign aid has gone to meet these needs. Sudan's population in 1985 was estimated to stand at 22·8m., having steadily expanded over the period 1970–82 at an average annual rate of 3·2 per cent. (Even the Ethiopian growth rate over the same period was reckoned to be lower.) There was accordingly increased pressure on diminishing water resources and overgrazing of available land, with further deterioration of the environment. To combat the threat of continuing desertification, Sudan received more support in 1985 from the West than any other African state. But even to complete a modest, nation-wide anti-desertification programme required a minimum $60m. supplement which, on past form, was not likely to be found.

The industrial north spends billions annually in unemployment benefits for people to spend their time in idleness, snatches away hundreds of millions more from higher education so that people cannot enhance their lives, and is apparently too jaded to see the possibility of rekindling the hope and imagination of its own anomic masses by engaging them in the Sahara and elsewhere in a common defence against the collapse of the globe's environment. It is no longer credible to go on moaning about 'wasting' money: the most vociferous complainants are those most adept at doing what they complain about. (Mr Reagan's 'Star Wars' initiative will license the outlay of countless billions on the pipe dream of a 'perfect' defence.)

Meaningful development aid was no more on the cards for Sudan than for Ethiopia, despite the former's supposedly friendlier (pro-Western) mien – at least under Nimeiri. By 1985, Sudan's accumulated debt of $11b. represented about $500 per head of population in a country whose per capita GNP in the same year was under $400. The interest alone on this astronomical sum would approximately equal Sudan's annual export earnings from all goods and services. But this sum was not fixed, it was growing. Sudan's debt in 1983 was 'only' $7b., and 80 per cent of foreign exchange earnings were then deployed to service it. The debt had been incurred in the first place in order to fund development, most ambitiously from 1972, and particularly in textiles and sugar. But foreign loans to sustain this effort evaporated within two years. In order not to drop what it had first begun, Sudan had to turn to the IMF in 1978. And the IMF imposed the shortsighted austerity measures that, by 1985, were universally associated with it: devaluation, cutting food subsidies, increasing cash crop exports, raising interest rates, lifting price controls, etc. There were the ensuing riots, military intervention to keep the lid on, and another change of regime, effectively introduced from without. We may still talk about Sudanese independence, but scarcely of Sudanese autonomy.

I have argued earlier in this chapter that the northern states, historically speaking, created the present-day southern economies. Northern creditors, through agencies like the IMF, merely ensured the continuing cheap supply of tropical products to temperate zones. Lowering the price of items like cocoa, cotton, coffee and tea in no way made these products more 'competitive'. Increased production in Sudan for export, joined with lower prices for these commodities, meant collectively more work and less pay. Of course attention is often drawn to the fact that in

Sudan, as in many African states, increased production for export, given that it does not exhaust all available arable land, cannot be seen as itself causing African famine. What is equally often overlooked is that, although increased export production at lower returns does not exhaust all arable land, it does absorb an increasingly higher percentage of labour, and this in itself will tend cumulatively to push a population closer to the edge of famine.

For each state in the Sahel, including nothern Sudan, there are variant local circumstances of which account is to be taken. The civil war in Sudan (between its north and south) has much to do with a criminal distraction of attention and resources from the famine of 1985. Mauritania's problems and Sudan's are not exactly the same, nor Mali's and Chad's. But all share increased production, decreasing returns, population expansion, greater pressure on land, decreasing aid, increasing militarization, often enough leading into war, within and between states – it is a story frequently re-told. Mauritania's problems differ from most of Sudan in detail, not in substance.

Mauritania[1] was once a land of endless forests, timeless caravans and hardened nomads who grazed vast herds for centuries on the extensive grasslands fringing the Sahara desert. A relentless tide of sand moving inexorably southwards has now largely swallowed the country. Many of its once fiercely independent herdsmen have become demoralized squatters in fetid refugee camps, surviving only on food hand-outs from foreign donors. A Western ambassador lamented: 'We are watching the death of a civilization and of a nation.'

One of a belt of states south of the Sahara which has suffered the tragedy of almost permanent drought and famine since the late 1960s and early 1970s, Mauritania has escaped the worst ravages of Ethiopia. Comparatively few lost their lives here. But, more than most states, Mauritania has suffered overwhelming physical devastation of the land and a social revolution among its 1·6m. people.

'We have no money. We have no animals. We have no rain,' said Mohammed Ould Elewa as he squatted under a tent of sheep's wool. A howling sandstorm sweeps across the landscape outside, reducing visibility to a few yards. There is no sign of movement in the camp from

1. The section on Mauritania (from here to the end of the chapter) was contributed by Raymond Wilkinson, *Newsweek*'s bureau chief in Nairobi.

its seventy-three families. Everyone squats and waits. There is no work here, no future. 'If we go to the cities our society will die,' he says. 'If we stay in the sand, we die. What can we do? We wait.' These people came here two years ago when their animals died. There is one well. Like 70 per cent of Mauritania's population, the nomads now survive on official hand-outs. If they are lucky, every two months, each tent receives fifty kilos of wheat, five kilos of milk and five kilos of butter.

Many villages in the interior contain no men at all. They have gone to the towns to find work or disappeared with their surviving livestock in search of pasture, often far away, into neighbouring countries. The women of the northern town of Rashid have not seen their menfolk or the herds for two years. Refugee camps have sprung up near large towns and along main highways and rivers. At one of these, Jeda Mint, Habibou Arahman and her sister study the Koran all day. It is their only pastime. There are no men in their village. 'We haven't seen rain for ten years. Our animals are dead. Perhaps God is punishing us,' one sister says. Suddenly, the two women laugh, a private joke. Why? 'We have become captives here,' one replies. 'But we are still alive.' Would they ever return to the sands? 'We want to build a home here, a big home to protect us from the wind . . . but' – a pause – 'but if we can get some new animals, we will go back. This is not life.'

Conditions are equally harsh in the towns. Nouakchott was built for a few thousand people at independence in 1960. By 1982, it had ballooned to 250,000 and will reach 600,000 by the end of 1985. The town's central core is surrounded by a sea of squalid tent cities and cardboard and tin shanties. It is reputedly the largest refugee camp in Africa. If they can find work at all, once proud nomads become nightwatchmen or occasional labourers. Their children become beggars or shoeshine boys. Most just sit and wait. They have become what one Western relief worker calls 'a colony of zombies'.

The social impact of all this upheaval is devastating. In the 1960s, three out of four children were born in the desert. This ratio has now been reversed. One child in five dies before the fifth birthday. One in three suffers from chronic malnutrition. Three children in four never go to school. The entire family structure is under assault. As women are abandoned by their menfolk in search of work, many turn to other 'husbands' or to outright prostitution. The UN children's organization, UNICEF, reports that, in trying to maintain themselves, 'some women have been married as many as twenty-four times, with serious consequences to their health. Others were grandmothers at the age of twenty-three.'

As a peace corps volunteer in the Mauritanian interior for several years, Mary Pecaut of Sioux Falls, Iowa, watched the disintegration of this ancient society from close quarters. 'One man lost his herds and simply went crazy because he could no longer support his family,' she recalled. 'There has been a large increase in crime. In many villages there are no men between the ages of eighteen and fifty. To support themselves, mothers sometimes "sell" their daughters for a night in return for tea or sugar.'

Dr Hussein Dia treats dozens of 'anxiety' cases at Nouakchott's main hospital. This type of complaint was unheard of until recently. 'Boys search for work in vain for years,' he says. 'They become depressed, crazy. Traditional values are under strain. We are in a period of crisis. Women come to me. They say they eat badly, see badly and have aches all over. It is not a physical problem. The real cause is deep anxiety.' He agrees that fifteen years ago there was little crime in this deeply Islamic nation, but now 'it has become a serious problem'.

The assault on the environment has been equally dramatic. The Sahara is moving southwards at five kilometres a year. In parts of Mauritania its advance is even more rapid. Four-fifths of the country's grasslands have disappeared in the last two decades. Part of this crisis was man-made. The ranges have been wildly overgrazed for decades. Even though herds have been reduced by 60 per cent in places during the current drought, there are still too many camels, goats and sheep for the fragile land to support. Forced from their traditional pastures, more than half the nation's nomads now live in shanty towns. Rainfall in 1984 was only 25 per cent of the recent average. The Senegal River, on the country's southern flank, is at its lowest since records began in 1904. One independent report said the average temperature had risen a staggering five degrees in recent times.

Some grazing land in the east could still be saved – with an investment of tens of millions of dollars and voluntary culling of the herds. But it is an unlikely prospect. The Sahara has already won most of the land. Sand dunes pile up ominously even in the heart of the capital. 'Where else do you have sand dunes in the middle of a capital city?' asks one incredulous diplomat. One of Mauritania's few paved roads is the 700-mile link running due east from the capital and optimistically named 'The road of hope'. Today it is lined with the tents of the world's latest refugees. For them it has become the 'road of despair'. Even here the sand is so pervasive that huge bulldozers are in daily action shovelling away dunes which cover the tarmac in a matter of hours.

A foreign ambassador remarked: 'Perhaps this is the only place in the world where people openly talk about the whole country just disappearing.' One UN report said that drought has 'increased dependence on free food, led to malnutrition and vulnerability to disease, internal migrations, abandonment of traditional nomadic pastoral life and urban congestion'. Last year Mauritania's farmers produced only 14,000 tons of cereal, less than 6 per cent of the country's needs. Currently the country has virtually no other income. Given these circumstances, the only future for these once-proud desert people seems to be to become international beggars in perpetuity.

# 6 Unnatural Causes

Drought, of the Mauritanian sort described by Ray Wilkinson, is itself an enemy to economy. But the severe effects of drought, taken together with a certain anticipation of its recurrence, always compel economy. Drought is not *ex nihilo*: if not strictly predictable, it is never altogether unexpected. People in drought zones devise arrangements to cope, or at least they try. Nomadic pastoralism, historically speaking, has certainly been one of the most brilliant of these arrangements, and one of the earliest. But modernity, and the centralizing governments that laud and promote it, have got in the way. The colonial administrations were generally pleased that they had created order, 'order' meaning the creation of a larger centre out of many smaller ones. Indigenous successor governments, like their colonial antecedents, were required to tap or generate resources to sustain this centre. The more barren the land, the more onerous the task. Centralizing governments like nomadic pastoralists least. They prefer the stable availability of settled agriculturalists. And the compact disposability of industrial workers represents, from a government perspective, the optimal arrangement. Colonial and developing governments generally prefer to encourage, even where ecologically inappropriate, tillage to pastoralism: it provides a surer source of revenue.

The colonial governments, while abolishing slavery, often enough resorted to forced labour as a means of implementing capital works or infrastructural maintenance programmes. A 'hut' or 'head' tax was often employed to achieve the same effect. But neither procedure was easy to apply to pastoralists. Indigenous tillers were excluded from the land, most especially in settler territories like Algeria, Kenya, Rhodesia and South Africa. But the people who suffered most, and who in the post-independence period continue to suffer most, are the erstwhile hunters and gatherers, and most importantly, the pastoralists, such as the Afar in Ethiopia or the Masai in Kenya. The agriculturalists expand and multiply. They gobble up grazing land previously available to pastoralists. These become restricted to proportionately less land, and to land still more barren. The colonial governments and their indigenous successors were and are bent on the cheap supply of metropolitan markets with a flood of agricultural produce such as coffee, tea, cocoa, bananas, sisal, cashew nuts and the like. It is an understandable strategy, but one which makes inroads upon food production for local consumption, not least among pastoralists. The produce of the land, transferred to the international marketplace, buys less and less: a process clinically formulated as 'deteriorating terms of exchange'. While the real value of African produce wilts, the defence and administrative costs of African states mushroom. The cheap (and effective) colonial umbrella against foreign dabbling in local affairs, long since whipped away, leaves the new men in the wet clasping a cardboard public purse that dissolves even as we watch.

African systems, being basically subsistence in type, have very limited reserves or surpluses. Independence placed greater demands upon those limited reserves. It was customarily accepted in the immediate post-independence period that such states should be assisted in their developmental goals by transfers to them from richer states. But not only did the volume of aid decline, in many cases it simply ceased. What happened in the 1980s 'debt crisis' was that the net outflow of aid and investment from the rich to the poor was simply reversed. Third World borrowers, including the African states, pay out to the First World huge sums in the form of interest, sums which exceed by far investment or assistance of any kind from the wealthy states. The IMF customarily encouraged African and other poor states to overcome the difficult position in which they found themselves by urging them to spend less and produce more. The trouble is that the primary goods produced by the Third World, on the whole, ran up against firmly inelastic demand: there was and is only so much gold, iron ore, copper, oil, copra and cloves that the rich world can consume. To produce more, paradoxically, simply meant that poor countries earned less: primary product prices collapsed. The effect? Poor states, with already exiguous resources, in attempting to build up their resources further through foreign exchange, fell further behind. Their resources have been diminishing, and an increasing percentage of what remains has been allocated to the escalating burden of military and administrative costs. Indeed, as the volume of revenue available to government falls, so the amount expended to contain public restiveness rises.

Drought is a factor, prolonged or not, that makes a dent in already limited reserves. But it means nothing on its own. It has to be taken as another critical blow to an already crumbling position. Drought may well be said to cause famine. But inadequate administrative resources may equally be said to cause drought. The inability to encourage and pay for family planning, to check population growth, to create jobs, to promote reafforestation and anti-erosion measures will all undermine a population's ability to thrive or even survive. Drought may well cause war, but war in turn may cause drought. Even where war does not do this, it will obviously deflect resources from defence against drought. War encourages combatants to use drought as a weapon. War is a much more potent cause of famine than drought. To understand large-scale human suffering, the rule of thumb is to look more to people than to 'nature' for the cause. Even where the human causes are not simple, and usually they are not, that is generally where the explanation lies.

It is no accident that the major theatres of famine in Africa, especially Ethiopia, Sudan and Somalia, have doubled as the major theatres of war. War has made planning impossible, even with the limited reserves in place, and has misdirected use of these reserves so as to damage and destroy infrastructure and services. Even Mauritania, now little affected by war, suffers its consequences: the misdirection of resources earlier forced upon this state in an unfortunate and prolonged joint venture with Morocco to appropriate Western Sahara proved an absolute disaster. In verdant settings, where famine should have no purchase whatever, it manages all the same to bite. South Africa extracted from Mozambique the 'Nkomati' peace accords, which accords, in league with dissident Portuguese elements, she systematically flouted, funding a continuing guerrilla war against Maputo, causing Mozambican famine. Severe drought tips the country over the edge; but without the war, which falls entirely within the province of man, the drought would have killed little more than time.

In Chad, civil war has raged continuously for years, with Libya backing various elements, appropriating a vast swathe of Chadian territory in the north and being fitfully opposed by successive French governments. The war has inhibited anti-erosion measures, disrupted production and forced populations into neighbouring states, not least into Sudan, with her already chronic problems.

Angola, harbouring the Swapo activists who struggle for Namibian independence from South African control, has been invaded again and again by South African forces. South Africa, burdened by its racist ideology, has sought to forestall Namibian independence, has seen the

Soviets and Cubans sucked into the region in defence of Angola, and in turn has successfully used the communist presence as a basis for US and Western support for her grandiose and strained ambitions for continuing regional dominance. When the MPLA government came to power in Angola in 1975, the USA was covertly disbursing $33m. to support the UNITA and FNLA oppositionists, all working in concert with the South Africans, whose armed forces had penetrated the country from the south by as much as 600 miles. Famine continued in Angola in 1985. Contributing to this, the US Congress proposed to prohibit new investment in Angola and to cut off $500m. of US investment already there, while simultaneously giving as much as $54m. of assistance (not least military) to the UNITA dissidents, all with a view to encouraging destabilization. South African bombing runs in September–October 1985 alone caused as much as $36m. of damage to Angolan installations. The Benguela railway has been repeatedly attacked, causing losses in the order of $60m. The Angolan government estimates that the country has suffered $10b. of damage to its infrastructure since independence.

The popular Western perception of this war, in so far as there is any perception of it at all, is as an irrational 'tribal' conflict. The trouble is that the major 'tribal' players are South Africa and the United States, and that without their active concern to break Angolan support for a non-racist South African future, there could not be the scale of suffering witnessed since independence, nor the famine which eventually evolved from such callous and fully self-conscious policies. It is not as though these policies are secret: the CIA clearly wished to break the government (MPLA) by supporting the rebel anti-communist front. Hence, too, subsequent American legislation to silence former agents who make such revelations.

In Uganda, the *coup* against the Obote government in 1972 was inspired and backed by Israeli and British agencies in hopes of defeating Obote's Common Man's Charter, which was regarded as threatening private Western, largely British, economic interests. The effect of this was to elevate the gross Amin, who repaid his foreign backers by expelling and imprisoning their citizens, but who did far worse by the Ugandans themselves, who were reduced to the most brutal dissidence and locked into a continuing civil war, with the deprivation and famine that mark such events.

To understand a phenomenon like famine, it is not enough to talk of drought. War is more important. Of the perhaps 125 wars that have occurred since the Second World War, over forty have taken place in twenty-five African countries – most of them civil wars. Statistically, every other African state has experienced this difficulty, but the seeds of such dissolution exist in all. Whether its reach is global or regional, a major power which wishes to exploit for its own purposes such a circumstance will not find the going hard. Rivalry between the great powers in particular, and especially in a region as sensitive as the Horn, swiftly becomes murder. The intensification of super-power rivalry is, in part, just another way of directly contributing to the killing of people whom the principals cannot claim as their own. It is now time to look at the play of war in a region of famine, and at the overlay there of inputs, local and foreign. Taken alone, war, like drought, will provide no complete explanation. But it should bring us somewhat closer.

# Part Two  WAR

# 7 Ramu Knights

The North-Eastern Province of Kenya is remote, wild, rockily barren, dry, indeed very much a desert land, with only a tiny population in relation to area. It is not a rich, green land, despite an overlay of cattle and sheep, but a bare brown land, ideally devised by God for camels and goats. It is a land of great distances, of small and dispersed populations, of sandvipers, of odd wells and random oases. It requires much patience to make a life in such a region. This patience rescues the nomad from extinction, but it is often joined with a rigidity which makes him its prisoner. He has not the air of one seeking to escape his condition, but of one who requires the harsh environment to justify his independence, his pride, which is unbending, hard, sometimes even pitiless, a pride which he has summoned up from some infinity of inner space to challenge the desert world which is his place.

The north-east of Kenya shares common borders with south-eastern Ethiopia and with western Somalia. These three states converge upon a singularly barren zone, which extends for perhaps 90,000 square miles, an area the size of the United Kingdom. There is no significant trade in, or between, any of the countries in this zone. Free movement is important for the pastoralists who inhabit the area, for they must search for water and pasture. They generate little revenue for their governments, and, from the latter, receive few services. There are no rail lines, or much of a road system. The economies of Ethiopia and Kenya, in normal times, are most vibrant in their highland centres; but there is little or no connection between the two economies at their peripheries. The major product of each country is coffee; and such products as this are shipped only eastward to the sea – to the Red Sea and the Gulf of Aden, in Ethiopia's case; to the Indian Ocean, in Kenya's.

Ethiopia and Kenya are dependent, in their trade with the outside world, upon access to the sea. In Kenya, that access is secured by a south-easterly road and rail link from Nairobi to Mombasa, both of which fall under the control of the central government in Nairobi. In Ethiopia, access is normally secured by north-easterly road and rail links to Asab and Asmara – which are in Eritrea (and thus legally in Ethiopia) – and to Djibouti, which became independent (of France) in June 1977.

To reach the sea from the fertile and productive areas centred on Addis Ababa and Nairobi, miles of scrub and desert must be crossed. These areas are peripheral to the centres, and certainly poorer than they are. Ethiopian and Kenyan road and rail lines run respectively to the north and south of their Somali neighbour; Mogadishu, the capital of Somalia, is also the capital of this coastal, intermediate and mostly desert land. It is its spokesman; it has a strategic position, and for the moment virtually no productive capacity. It seeks a role for itself, a role that can be bought only at the expense of the highland and hinterland which it borders, and whose access to the sea it inevitably seeks to control and, in various fashions, to tax.

Journalists and laymen tirelessly inquire why it is that anyone should covet such a wilderness as the Horn of Africa. There is of course the strategic importance of the area for Western and even Soviet shipping. There is the probability, further, that great mineral wealth lies hidden underground, and that this is known in domestic and foreign chancelleries. Finally, and paradoxically, an excess of barrenness may serve as a condition for covetousness. For there is so little pasture and water that what there is constitutes a boon intensely competed for. No stretch of this land is so fertile that it will produce what the settler requires for survival. The land sets its face against agriculturalists. It smiles only upon those who do not scar it with digging and tilling; it welcomes the wanderer.

In June 1977 the Kenya government was thrown into a panic by the prospect of a sudden and violent loss of its north-eastern corner, constituting a fifth of the country's entire land area. The suspected invaders were the nomadic pastoralists who inhabit the area, together with regular units of the Somali army, whose government covets it. Significant violence had erupted in Kenya's North-Eastern Province, and senior officials were persuaded that an undeclared war of major proportions was being forced upon them. They were in great and justifiable doubt about their ability to defend against the suspected assault.

We have no official account of what really happened, and can only attempt a reconstruction. By the beginning of June 1977, large numbers of armed Somalis had massed at two distinct points (at least) along Kenya's north-eastern border with Somalia. Nairobi received reports of these developments, fully confirmed by the second half of June. There was uncertainty about the actual number of Somalis involved. The Kenyan government suspected – probably in June and certainly by July – that as many as 15,000 either were or would be assembled along the border for purposes of infiltration into Kenya. The entire Kenyan army was only about 8,000 strong and was not in any case greatly trusted by the civilian authorities. In June, Kenya had only about twelve jet fighters of British manufacture, and no genuine means of securing air superiority over the Somalis (with fifty Migs). It seemed certain that Kenyan forces could not withstand an attack, given a combination of restiveness among the native populations of the north-east, together with the overwhelming numbers, modern weapons and potential air support of the prospective invaders.

Two groups of invaders, by 23 June, were known to be assembled on the Somali side of the border – they were too large to escape detection, even in the wilderness of the north-east. The Kenyan army and police were ordered into position on Kenyan soil opposite the invaders. Their numbers were small (a company) in relation to the size of the first group of invaders, estimated at 3,000 men. The Kenyans had reason to expect serious trouble, but hoped to avoid it by intercepting the Somali column on Kenyan territory and turning it back. On 24 June, the first Somali column pushed west into Kenya, setting out from a point roughly thirty-eight miles south of the town of Mandera. The company of Kenyan army and police intercepted the column at Malka Figo, six miles west of Mandera. There was an intense exchange of fire. The Kenyans were overwhelmed and swept aside.

On the next day, 25 June, a second and larger armed group of about 3,500 men crossed into Kenya from Somalia and quickly converged upon Malka Figo. This second group was not interfered with by the Kenyans, but was merely observed from a respectful distance as it moved at a shallow angle from south-east to north-west, towards the Kenya–Ethiopian border. On 26 June, the two columns of invaders, which had entered Kenya separately, converged somewhere east of the Kenyan town of Kalalio, forming a powerful force of 6,500 men, still moving west towards the Kenya–Ethiopian border. The Kenyans were now confirmed in their view that the purpose of the invaders was to absorb Kenyan territory. The invaders were advancing upon the Kenyan police post at Ramu and the nearby Ethiopian post at Sadei.

On 27 June, the combined Somali column reached Ramu and Sadei.

The Somalis fell upon both the Kenyans and the Ethiopians, and overran both posts. The airstrip was mined to prevent reinforcement by air, as were the roads to block overland reinforcements. Official Nairobi was shaken by the news, even if it did not immediately reveal its discomfiture, or ever reveal in detail the true reasons for it. In fact, one of the most interesting aspects of the Ramu incident was the disinclination of the Kenya government to say very much about it, save to protest that it had occurred and to warn the Somalis of retribution if such behaviour continued. Kenya, indeed, handled the incident in a strikingly circumspect fashion. No public word was leaked about any of these clashes until 29 June, five days after the first had occurred and two days after the last. The press reports that then emerged still made little sense. No clear and full account was ever provided – either to the public at large or to the Kenyan parliament.

Kenya was embarrassed and perturbed by these attacks. She took her time in devising a public stance, but finally decided to hold Somalia responsible without saying precisely what had taken place. On the contrary, it appears that the evidence was reconstructed so as to cover up the seriousness of the situation. In the pre-Ramu clash of 24 June at Malka Figo, the Kenyans claimed that six of their men were killed, that seven were wounded and that they lost only a few weapons; they also maintained that six of the invaders were killed and thirty-five wounded. The Kenyans further maintained that on 25 June, the day following the clash at Malka Figo, another group of 3,500 armed men 'appeared in the same area and after pursuit by Kenya armed forces they fled into Ethiopia'.

As to the clash of 24 June, the Kenyan account does not make sense for several reasons. It is unlikely that a mere company of soldiers, lightly armed, would seriously oppose a force of 3,000 men, armed with sophisticated weapons. If they did do this in fairly open country they could easily have been surrounded and wiped out. But if they were not, it remains improbable that the number of dead or wounded among the larger force would be equal to or greater than the dead and wounded among the smaller force. Suppose the Kenyans actually to have delivered as lethal a blow as their figures imply: it is none the less the Kenyans, by their own admission, who were compelled to withdraw. And in doing so, it is difficult to imagine that they would have had the leisure to undertake a body-count. The conclusion that there is something odd about the Kenyan version is not easy to escape. But what its oddity does not bring into doubt is the fact that Kenyan units were badly mauled by alien forces in action within Kenya on 24 and 27 June at Malka Figo and at Ramu respectively.

The Kenyans will have had little or no idea of the damage – if any – they inflicted upon the invaders; their figures for enemy dead and wounded must be ignored. They were probably concocted so as to enhance the image of the Kenyan forces. As for 'pursuing' the invaders – who according to reports constituted a force almost as large as the entire Kenyan army – it is more likely (and prudent) that a mere company, already sorely tried, would have reconciled itself to enjoying the role of an observer. There was no need for the invaders to 'flee' into Ethiopia; they were moving into Ethiopia anyway, with such speed as the terrain and their transport would allow. Their direction was fixed before they crossed into Kenya, and maintained in total disregard for the violation of Kenyan territory and in defiance of the Kenyan armed presence.

There is no evidence of any Somali wish to avoid the Kenyans. If anything it appears probable that they deliberately sought an early armed confrontation with them to dissuade them from future intervention. The immediate objective of the Somali invaders was less to appropriate Kenyan territory than to destabilize Kenyan administration and, by crossing Kenyan territory, to achieve early and easy access to the soft Ethiopian underbelly at those points where the latter country was least well defended.

There were deaths at Ramu. There was movement across borders, too. But then the determination of a border in a desert is as meaningless to its people as nets of cotton to sharks at sea. None the less, something new was involved. First, these attackers did not come in small clusters. They came in force, with all the prior coordination, planning and training that this must require. They were not mere rustlers and poachers, although their training did not disallow a bit of either. Their rifles were not the rusty old kind with home-made ammunition to which travellers in the area are accustomed; instead, they were equipped with modern automatic weapons of the sort issued to the Somali army. They did not seek to avoid conflict with the regular forces in Kenya; rather, they neatly disposed of them, partly because that was the most efficient way of removing an obstacle, but partly also in order to teach the weak their place in a world dominated by the strong. The invaders were not casuals; they had the training of regular soldiers under the auspices of the Somali Democratic Republic.

There was little the Kenyan authorities could do, except to protest to the Somalis, to pray that this attack was not what it seemed, to hope that the Somalis might be chastened by appeals to the OAU, and/or that friendly Western powers would come to the rescue. The Kenyan protest to Somalia generated only a bland denial of responsibility and the allegation that this devilish assault had been mounted by the Ethiopians. The Kenyans raised the matter at the OAU summit held in July. And the new Ethiopian head of state, understandably, condemned the Somalis in far more sweeping terms.

From June 1977 onwards Somalis, of whatever 'nationality', circulated at will, back and forth, within and beyond Kenya's borders. Where there were cattle and camels and game, of which they might make use, they killed or took them. Where there were officials or individuals whom they opposed, they grew accustomed to eliminating them. They attacked the army, the police, villages, outposts, individuals and game wardens. And the Kenyans did not dare bear down on them too hard, for fear of reprisals.

The Somalis made strenuous attempts to calm Kenyan fears. President Siad Barre dispatched his Vice-President, Hussein Afrah, to plead the case, with a personal message for Kenyatta, which the latter refused to receive. The Kenyans publicly vented their anger while carefully avoiding any provocation of their adversary, by military or judicial means. (A band of heavily armed Somali soldiers, apprehended on Kenyan soil, were quickly released to the Somalis after twenty-four hours.) Siad Barre, from Mogadishu, obliquely warned neighbouring states (meaning Kenya) that intervention would earn them a severe thrashing. The Kenyans, behind their complaints, kept quiet about the extent to which security in their North-Eastern Province had deteriorated.

The Somali government was plainly concerned to have Kenya believe that the latter was not under attack, and the Kenya government was relieved to believe that this was so. But Kenyan officials perceived, all the same, that a severe blow has been dealt them, and that the Somalis had resorted to semi-placatory gestures only in order to avoid an untimely and large-scale confrontation of a public and international kind.

Throughout the second half of 1977 and into 1978, the Somalis were freely crossing into Kenyan territory, using Kenyan livestock for their catering and Kenyan manpower for incorporation into their military.

Schoolboys were disappearing across the border, often no doubt voluntarily, but often not. It came to light that the Somali embassy in Nairobi was regularly and fairly openly issuing Somali passports to Kenyan Somalis to facilitate their travel to Somalia for further military training. The Somalis admitted issuing passports, but denied that the purpose was military. In their obstinately warm and congenial fashion, they insisted that these passports were intended to allow travel into the Arabian peninsula for work purposes, and that their involvement was merely humanitarian. The Kenyans suggested that as many as 2,000 passports could not have been issued on such grounds. The Somalis retorted that only 200 were issued. A Kenyan immigration department spokesman pointed out that hundreds of Kenyan passports had been issued to Kenyan citizens to enable them to take up work in the oil-producing states, and insisted that Somali Kenyans encountered no difficulty in obtaining passports for this purpose. Of course, Somali applicants *did* encounter difficulties. What stands out from these exchanges is the Kenyan conclusion that she was, yet again, being put upon.

By the beginning of 1978 there was no doubt left that the Somalis were operating freely within Kenya's territory and that there was very little the Kenyans could do to stop them. Clashes occurred far more frequently than was reported. Even when Kenya did report such incidents the tendency was to play them down as 'banditry' and the like. *Matériel* and personnel were freely moving from Somalia, through Kenya, and into Ethiopia. In the immediate period of the Ramu incident, the Kenyans sustained a direct Somali attack. Kenyan officials at first thought that the Somalis wanted territory. Within a few weeks they recognized that Somalia's immediate aim was less to absorb North-Eastern Kenya than to make free use of it, without ever publicly acknowledging their official involvement. From this point, Kenya was persuaded, both by foreign powers, and out of domestic considerations, to calm down. She found it more politic to project an image of disinterested support for the Ethiopian position, without caring or daring to divulge just how thoroughly she had been beaten by the Somalis and just how hopelessly vulnerable, in 1978, she still remained.

Nevertheless, more and more information, albeit confused, slipped out. In December 1977 there were reports of raids which extended from the south of North-Eastern Province westwards through Isiolo, the Tana River and Kitui. In one of these, hundreds of cattle were stolen and a man killed. In a second, in the same month, several hundreds more cattle were taken and yet another man killed. There were reports of fleeting contact between police and 'rustlers'. On 24 January 1978, there was another raid on a little town called Sololo, just three miles from the Ethiopian border, in Kenya's Marsabit District. Twenty-three Kenyans were killed; the same number needed hospital treatment; and this was followed by reports, as at Ramu, of Kenyan units rushing to the area and of the raiders 'escaping' across the mountains into Ethiopia.

In most cases, the raiders were said to be Somali. It was claimed that, by February 1978, they had killed 100 Kenyans. The Member of Parliament for Isiolo North, A. H. Fayo, stated that bands of Somali, in groups of up to fifty, armed with automatic weapons, were entering Kenya almost daily, both for raiding livestock and for transit purposes. But, in early February, just before Fayo made his statement, a former member of parliament from the border area, Osman Araru, appealed to the Kenya government to end 'the killing of innocent people' in his area. He maintained that there were 'tribal wars' going on in his native Marsabit. He described a particular attack, one of many such, on a Galla (more specifically, Boran) camp, which he dated roughly as taking place on 12 January. By his account, twenty people were killed in this raid, fifteen wounded and 2,000 head of cattle taken. Interestingly, Araru, himself a Boran, did not accuse the Somalis. He accused other Kenyans, namely the Garba and Gari peoples. Moreover, he maintained that the latter were only retaliating for what the Boran had earlier done to them. He thought the cycle would continue, particularly in circumstances where administrative officials were overly disposed to accept the accounts put to them by the Boran chiefs (to the effect that those responsible were Somali *shifta*, or bandits). Araru firmly insisted that the Somali were not involved at all.

Araru's statement, made at a press conference, put the Kenya government in a ticklish position. If the government confirmed Araru's version of these events, it would be admitting an utter lack of control over almost the whole of the northern and north-eastern border area. If it denied Araru's version, it would be admitting, more plainly than before, an utter inability to defend its people against attack from neighbouring states. A fairly low-level official in the Ministry of Information and Broadcasting, on the day following Araru's bombshell statement, declared it to be false, and insisted that Somali raiders, armed by the government, were involved.

It is fairly clear that the Somalis were responsible for much mischief. For such mischief as they did not cause, they served as marvellously convenient scapegoats. In fact the Kenyans increasingly blamed them for all manner of evil, such as the killing of civilians and park rangers and even (occasionally) a policeman or two. But what could not be publicly revealed and openly recounted was the chiefest and most central of these evils, that which really explained the occurrence of all the rest: that the Kenyan army had been beaten and would be beaten, and was not even in the ring for the count.

The rout of the Kenyan army, which was withdrawn from the area, caused a state of insecurity throughout a 'border' region which, at the extreme, may have equalled half the area of the country. Many of Kenya's provincial administrators were rendered virtually powerless. There was no way of protecting Kenyans from the Somalis, but this meant that there might be no way in which Kenyans could be protected from one another either. A raid may be mounted on a neighbouring people's herd, after the manner of the Somali *shifta*. There might be dead and wounded. There was no authority to stop it – the Somalis could always be blamed.

It was easiest for the Kenyan government to do nothing, apart from indulging in verbal protests to the Somalis. If they pursued the *shiftas*, the army would be ruined; if the army won, the government itself might be under threat – and so itself be ruined. There was no rush, once it became clear that the Somalis were not making an immediate grab for land. Kenya would indeed have to build up her armed forces, but slowly, surely, with full Western (American) backing, and it was especially important to build up her air force. Three-quarters of the country was not crucial to its economy anyway, nor therefore to the governmental elite. It was the south-western quadrant – populous, productive and secure – that really mattered, not the rest. What Osman Araru may in part have been getting at was not that the government was powerless, although this was certainly a large part of the truth, but that it did not find it prudent to exert such power as it had. Nairobi, after all, spent precious little on any area outside the south-west. And if those scattered and mostly pastoral folk should choose to blood themselves, was it wise, whether economically, militarily or politically, to intervene?

Next to protesting to Somalia, to the OAU, being nice to the Soviets, and entreating friendly Western powers not to arm or provision Somalia, Kenya increasingly became more concerned to whip the Somali community in Kenya into line, at first discreetly, and then more forcefully. The apparent object was to cut off the Somali raiders from any ethnic support they enjoyed within Kenya – whence they regularly received supplies of food. Restrictions were imposed on the movement of goods by car, camel or otherwise across the border. Such action would not stop the movement of insurgents. It imposed some hardship on local traders, who were further threatened with severe punishment if caught smuggling across the borders with Somalia and Ethiopia. By early 1978, many Kenyan Somalis – functionaries, chiefs, traders – had been gaoled indefinitely and without trial because they were considered to be security risks.

Again, this constituted an easy way out for the government. The harassment of Kenyan Somalis in the border areas was followed up by tightening the screws on the Kenyan Somali leaders. There were rumours of Kenya government officials – Somali ethnics – clandestinely provisioning the raiders across the border, of their 'sowing seeds of discord' among the very people they were meant to be rallying to Kenya, of their being devoid of committed leadership, of their not playing their role to bring the Somali nomads on side. At the same time, plans were being laid, as announced by Kenya's then Vice-President, Mr Daniel Arap Moi, to compel citizens to carry identification cards. The measure was ostensibly intended to cover all Kenyans; but it was planned to begin in the North-Eastern Province; it might just happen that it would not be extended elsewhere. It was clear in any event that the Somalis were the target community. Members of parliament from the border areas were attacked in statements leaked to the local press and attributed to 'informed sources'. The assumption appeared to be that, if there were problems among the Somali nomads, it was the Somali representatives in the Kenya government and parliament who should be held responsible. And the test of their loyalty was made to consist in how far they would go in denouncing the Somali menace.

Of course, this was an impossible position for any ethnic leadership to be put in. The Kenya government in fact and understandably did very little for these border areas; it would be easy always to do more, but difficult, for a variety of considerations, to do a great deal more. Parliamentarians from these areas exercised no power in Nairobi, and could bring back to their home districts very few benefits. They desired, naturally, the support of their constituents. Their constituents, often and volubly, had reason to doubt that such loyalty was at all reciprocal.

Many Somalis were disenchanted with the central government, but most particularly with their own representatives there. It would not be a difficult matter, in such circumstances, for almost anyone, including agents of the government of Somalia itself, to stir up trouble by promising local people a better and more prosperous life should they revolt against Kenya (or assist others in doing so). The last place for the Somali member of parliament to be, with nothing to offer, was in his constituency. There he would be held accountable as a servant of Nairobi. Yet in Nairobi, it was he – not the officials of the provincial administration effectively governing – who was being held increasingly responsible for untoward developments in his constituency. It is the sort of situation calculated to produce the type of 'representative' whom both centre and periphery can call Quisling, and whom neither can trust.

The night of 27 June 1977, at Ramu, was not a glorious one for the lightly clad Kenyan military. It was one of those drubbings which the government had to own up to in order to be able to complain about it – both to Somalia and to the international community at large. But in fact it was only a single incident, forming a part of a larger pattern of events which had been building up in the Horn of Africa for a long time. It provides as good a picture as any of how Somalia initiated a war, principally against Ethiopia, in the course of which tens of thousands of humans were killed, hundreds of thousands displaced from the war zone, vast populations of zebra, cheetah and other animal species indiscriminately slaughtered. Such mayhem is nowhere near its term. One must probe well beneath the surface immediacy of obvious idiocy to make any sense of it.

# 8  Eritrean Flashback

By the beginning of 1977, the civil war in Eritrea had become critical for the Ethiopians. Given the internal change of regime, the Western (mainly US) flow of arms into Addis Ababa had slackened. Addis Ababa's final request to the Americans for $65m. worth of ammunition was refused. On 11 February 1977, it was announced that the Ethiopians would look to the Soviets for military supplies. On 24 February the USA responded by cutting back on military aid and credits to Ethiopia and protesting about human rights violations in the country. On the matter of violation of rights, Ethiopian radio (on 24 April) countered, arguing that the USA had not withdrawn aid from the late Emperor, even though it was widely known that he was responsible for 'decimating thousands' of oppressed Ethiopians. On 23 April, in response to the American cutbacks, Addis Ababa expelled the US Military Assistance and Advisory Group and the US Naval Medical Research Unit and closed (along with other centres) the US communications facility at Kagnew in Eritrea. The facility at Kagnew was, in any event, being phased out because of the increasing use of satellites. On 27 April, the USA simply suspended all further arms supplies. From this point the Ethiopians were fully dependent, militarily, on the Soviets.

Largely because of their alliance with Somalia, the Soviets were not as forthcoming as the Ethiopians wished or required. Nor did the Ethiopians wish to become as dependent as they were destined to be. They desperately needed to suppress internal dissidence in general but were especially anxious to secure Soviet support in facing up to the Eritrean secessionist challenge. To do both, it seemed essential to avoid a brawl with the Somalis, whose tank force had become the most formidable in the whole of eastern Africa. The Somalis, by contrast, encouraged matters to move in their direction, but both Somalia and Ethiopia maintained mutual diplomatic relations and neither was against discussion. Over the period 14–16 March 1977, Fidel Castro made a secret trip to Ethiopia. A short, discreet meeting was arranged between Mengistu and Barre in Aden. There Mengistu proposed some form of federation (Soviet-inspired) between South Yemen, Somalia and Ethiopia.

President Siad Barre's concern was to acquire the Somali-speaking areas of southern Ethiopia, basically Bale and Hararghe provinces: the Ogaden. Mengistu could appear to agree to concede the Ogaden to Somalia only if such a cession was designed so as to sidestep any suggestion of an irreversible loss of sovereignty. A scheme was therefore advanced to the effect that Somalia should agree to a federal union of the states of the Horn; that Djibouti's sovereignty, after independence on 27 June, should be underwritten by an external power (to prevent Somalia taking it over and closing off the line to Addis); and that Somalia should cease to support the guerrilla movements which she had massively re-supplied and extensively re-organized over the preceding months. Siad Barre was not disposed to agree to any of these conditions.

The common ends both of the Eritreans and of Somalia could not be served by Somalia joining a union with Ethiopia. Mengistu (not unlike the USSR and Cuba) was merely clutching at straws. But he had little option. Otherwise he would have to swallow the bitter pill of a two-front war, a war already complex enough. The Soviets, naturally, were anxious to forestall an outright contest between their clients in the region, a contest originally made possible only through their foolish and self-interested supply of arms to Somalia in the first place. Neither Ethiopia nor Somalia could easily ignore a request by their Soviet patron at least to parley.

The Eritrean guerrilla movement (with an estimated total of about 37,000 men) had grown swiftly. It was fully supported, financially, by states like Iraq, Syria, Kuwait, Saudi Arabia, Qatar and Abu Dhabi. It was supported strategically and politically by states like Sudan and Somalia. The Ethiopian army had been beaten in a number of engagements. On 6 April 1977, they lost Tessenai near the border, on the main road into Sudan, permitting the guerrillas to re-supply by land and thus with ease. The Eritrean Liberation Movements were in control of a number of towns and were strongly deployed over virtually all other important population centres in the province. The Eritreans represented a formidable force deployed in equally formidable terrain and threatened not only a loss of real estate, but secure and sovereign control of a corridor to the sea.

Ethiopian forces were not only besieged in the north by the Eritreans. Their communications with the south were interrupted by other guerrilla units as well. The two main roads from Eritrea south to Addis Ababa lead in the west through Begender, and in the east through Tigray provinces. Both were cut. The road through Gondar in Begender was cut by the Ethiopian Democratic Union (EDU), with 10,000 to 15,000 men under arms. On 5 April 1977, the EDU captured the town of Metema. Metema, like Tessenai, lies on the Ethiopian border with Sudan (directly opposite the Sudanese town of Gallabet). The capture of Metema by the EDU, like the capture of Tessenai by the ELF, opened up for the insurgents a direct supply route into Sudan. On the eastern side of the plateau, a little over 500 kilometres (300 miles) away, the road leading through Makele, capital of Tigray province, was also cut. In this way, Eritrea was sealed off by land.

The military situation within Ethiopia grew worse, hand in hand with increasing uncertainty at the centre regarding what should be done about it. As the military situation in Eritrea, in particular, deteriorated, so did the serenity of the Dergue. Mengistu and others were determined to hold Eritrea. At the end of January 1977, Brigadier-General Teferi Banti, as Chairman of the Dergue, publicly appealed for support for the faction seeking a compromise with the Eritreans. On 3 February 1977, without the consent of the Vice-Chairman of the Dergue (Lieut.-Colonel Atnafu Abate), Mengistu arrested the Chairman and his associates, called a full meeting of the PMAC, submitted evidence against these men and had them executed posthaste. The officer who carried out the executions, Major Yohannis, apparently believed that he himself was to be similarly dispatched. There ensued a shoot-out. The upshot was that five security men were hit, including two of Mengistu's bodyguards, four of whom were killed, leaving Mengistu unscathed. It was following these events of 11 February 1977 that Mengistu became Head of State with Abate as Vice-Chairman.

After the failure of Mengistu's attempt to placate Siad Barre in March in Aden, the attention of the Ethiopian centre was focused more exclusively on the Soviet Union. A peasant militia was being formed to cope with the Eritrean problem. It needed to be trained and armed. The task was all the more urgent in view of the Somali attitude. Mengistu, accordingly, undertook a five-day visit to the USSR (4–8 May 1977). This was not the first visit of an Ethiopian leader to the Soviet Union.

The erstwhile chairman of the Dergue, Banti, had gone there in January 1976; Haile Selassie had also visited the USSR, first in 1959, then again in 1967 and 1970. In short, the Ethiopian leadership, before and after the 1974 *coup*, was not disposed to put all its eggs into one basket. The Mengistu trip was a very delicate one. He is bound to have raised with the Soviets the question of arms on the one hand and the question of Somali intentions on the other. Yet the joint communiqué issued at the end of the visit made no reference to military aid of any description – nor even to Somalia. The May 1977 trip was Mengistu's first excursion abroad since hacking his way to power three months before. In a speech delivered in Moscow, Mengistu spoke scathingly of the USA and of Sudan. Yet again, Mengistu made no mention of Somalia. The Soviets continued to signal their reluctance fully to re-equip the Ethiopian army. They did not wish to alienate the Somalis, or the Ethiopians to alienate the Soviets.

The Soviets hoped to offer sufficient arms to fix the Ethiopians without causing the Somalis to bolt. Ethiopia could live with the lower level of Soviet aid for as long as she was certain that the Soviets were in a position to restrain the Somalis. The Soviets obtained from Somalia assurances that they would not invade Ethiopia. The Ethiopians, quite specifically, sought to re-deploy some of their troops from south to north to contain the growing menace of the EDU in Begender. It appears that the Soviets were able to persuade the Ethiopians that the Somalis could and would be kept under restraint. One of the brigades of the Ethiopian 3rd division, headquartered at Harar, with overall responsibility for Hararghe Province, was later relocated to the north, significantly weakening the defence of the Ogaden. Soviet 'control' over the Somalis, when the crunch came, proved to be effectively non-existent. The Somalis did not slip the leash: there was no leash.

The Eritrean problem remains a running sore for Ethiopia. There will be no modernization of the country until it is resolved. In essence, it is a simple matter. Ethiopia has historically always been attacked from the sea. Italy controlled Eritrea in 1898 and in 1935, attacking the highland empire from coastal colonies. With other powers controlling her coastal outlets, Ethiopia felt herself to be, and indeed was, vulnerable. After the Second World War, Eritrea was not granted independence but was attached to Ethiopia (1952) on a federal basis. The Emperor later cajoled Eritrea to abdicate from its federated status, and then swallowed it whole (1962). The Italians had ruled Eritrea in the abusive and harsh racist fashion of the time. On the other hand, Italy had brought her into closer touch with developments in the modern world, and after the war had been allowed to prepare the territory for independence. In every important sense, Eritrea, in 1952, was rather more advanced than Ethiopia, both materially, and regarding democratic procedure. There was internal Eritrean opposition to an Ethiopian merger from the start, and it has never since ceased.

Guerrilla opposition to Addis Ababa was not slow to begin. The difficulty was for Eritrean dissidents to locate effective allies. They originally found support in Nasser's Egypt, allied with the Soviets. By the same token they found no sympathy in the USA, which at first discouraged support for Eritrea from Arab allies across the Red Sea. With the change of regime in Addis Ababa (1974), and even before, there were suggestions that American policy might be better served by reducing support for Ethiopia, with a view to improving rapport with Arab clients. In any event, with the decline of US support, the serious-ness of Eritrean dissidence mounted. But at the same time, while the Eritrean revolt was initially against Ethiopian imperial autocracy, the 1974 change of regime held out, at least formally, the prospect that significant social reforms would ensue.

Eritrean secessionism, like all such expressions of dissidence, follows on from acutely felt injustices. The new Ethiopian regime claimed that a solution, given the change of regime, was now possible. The secessionists were locked into their programme and could not readily back away from it. Attempts to negotiate had failed, and were probably not very serious from the outset. The Eritreans sought full independence or nothing. The imperial autocracy had always resolutely spurned any solution vaguely approximating to such an outcome. The Dergue, on acceding to power, and despite a degree of stalling on both sides, equally balked at so radical an innovation. The Dergue began its life as a genuinely elective organization; it awaited a major shake-out. The major force within the Dergue, outraged increasingly by their picture of Eritrean hostility, resistance, sabotage and indeed treason, began firmly to embrace the view that the sole effective means of dealing with an element that had declared itself an enemy was by resorting to untempered reliance upon sword and fire. The Eritrean leadership, equally outraged by Ethiopian air and artillery strikes, by arrests and executions, were reinforced in their view that the struggle must be intensified until either death or liberty was won.

Most of Ethiopia's Arab League neighbours – Somalia, Djibouti, Sudan, North Yemen and Saudi Arabia – were and are disposed to entertain a very lively sympathy for the Eritreans. Such sympathies have little to do with left/right divisions, and more to do with habit, religious disposition and common cause. Ethiopia, perceived as a threat to her immediate northern and eastern neighbours, was far from being an object of adoration. Her neighbours were largely inspired by the fairly common Arab urge to compel cultural, religious and strategic criteria to converge in structuring and re-structuring state systems. Every state, at many points, operates unjustly, and in this has much to answer for. But no state is uniformly opposed by other states merely because of its unjust practices. International morality tends to be highly selective. And so it is with the generalized Arab opposition to Ethiopia's handling of Eritrea.

With the increasing alienation of the USA from Ethiopia, much more effective assistance to the Eritreans came from neighbouring states sympathetic to their cause. Eritrea, with the exception of the southern end of it (Assab), was, by 1977, effectively cut off from the rest of Ethiopia. The port of Massawa could not be used. The guerrillas intensified their campaign in the province. In early 1977, the Sudanese banned Ethiopian overflights, including – most importantly – Ethiopian Airways connections with Europe. Significant guerrilla elements, introduced via Somalia, had been in place in the south-eastern stretches of Ethiopia since the previous year. The Ethiopian heartland, which does not significantly communicate with the external world via Massawa, was compelled to recognize its absolute dependence upon the Rift Valley route, by road to Assab, and by road and rail to Djibouti; of these the more important connection was the rail link. The railway was an obvious target. It had been attacked before. But to make these attacks good, one would have not only to destroy sections of the line, but also to be able to hold adjacent terrain, to exclude any prospect of repairs. The line could be firmly severed only as part of a broader coordinated effort to dismantle the Ethiopian centre. It was against this backdrop that a crucial struggle began in late May 1977 when the Addis–Djibouti line was blown.

In early 1977, Ethiopia was caught between the demise of the American alliance and the clear emergence of the Soviets as substitutes. In these circumstances, the obvious vulnerability of the Ethiopians served as a spur to Somali ambition. If an enemy were ever to entertain any reasonable hope of successfully detaching coastal and southern Ethiopia from Addis Ababa, then May, June and July 1977 presented an ideal time to strike – swiftly – before Ethiopia's fortunes improved.

Small powers do not normally like large powers for neighbours. They may bring comfort, but more often they are perceived as representing a threat to which there is no credible response. Yugoslavia (and Western Europe more generally) *vis-à-vis* the Soviet Union, Papua New Guinea *vis-à-vis* Indonesia and Mexico *vis-à-vis* the USA are examples. Ethiopia, naturally enough, would be delighted to live in a world where she directly controlled all the peripheral states with whom she must traffic in the Gulf of Aden, the Red Sea and the Indian Ocean. And these would be delighted to be able to contrive a situation in which Ethiopia existed (if at all) only in a much weakened and diminished form. Ethiopia is the source not only of the Blue Nile (so that Sudan is utterly dependent for water on her), but of virtually all the drainage into southern Somalia, including the Juba and Shebelle rivers, which are Somalia's only major rivers and which are vital to the economic life of the south. To remove the threat that Ethiopia might be perceived to represent (due to her geographic location, drainage, comparative fertility and population size) it would not be enough to appropriate the periphery. One would have to entirely dismember the state.

Somalia could not think of controlling the lowlands of southern Ethiopia, particularly Hararghe, Bale and Sidamo Provinces, unless, whether alone or in alliance with others, she was able to restrain or destroy the Ethiopian centre. Similarly with the Eritreans. It would appear singularly difficult for them to throw off the yoke of the Ethiopians unless Ethiopia were hopelessly dismembered, which would at least require Somali control of the Ahmar Mountain chain and of the Ethiopian plateau south of the Rift Valley.

# 9  Somalia: State of War

Somalia has a small population, 75 per cent pastoral, dispersed over a vast area and subsistence in character. The government has difficulty maintaining any continuous contact with it, is normally able to offer it little and derives equally little from it. Government revenue accrues from a tiny number of exports: livestock, fish and bananas. The life of a Somali, with respect to the central government, is thus normally a very free one. The central government normally impinges on the day-to-day activities of the average, and especially the pastoral, Somali to a negligible degree and is controlled by its subjects to an equally negligible degree.

Within the sphere of its operations, however restricted, the government enjoys a considerable freedom of its own, particularly in its relations with foreign aid donors. Between 1963 and 1977, the most important of these was the Soviet Union; subsequently it was the United States. The Somali pastoralist enjoys the freedom which is a prerequisite of his life-style. The small Somali middle class enjoys considerable advantages as well. But the incoherence of the Somali people as a political force imposes upon the Somali government severe limitations in its ability to resist or respond to 'popular' inclinations. The common interests of the populations represented are so brittle that there is little foundation on which the parliamentary procedures of a highly industrial system might be built. So a state like Somalia appears almost inevitably autocratic. There is no evidence that to devolve this state into a host of tinier constituent parts would in any way improve the hopes or fortunes of its peoples. Thus Somalia's relaxed military autocracy becomes the mirror image of the pastoral independence of its citizenry. The *de iure* power of the government, and the *de facto* powerlessness of the nation, are as one.

The structure of the Somali government is such as to confer upon the central government considerable independence in foreign policy. Lacking a substantial and reliable internal tax base, it is naturally tempted to turn foreign policy itself into a substitute tax base – or at least into a major source of revenue. Where Somalia succeeds in

international diplomacy, she does two things: she attracts foreign assistance, and she deflects the wrath of a volatile, domestic constituency.
War is one of the most potent means by which the Somali government
can discipline its own people, recruit them, contain them, offer a hope
consistent with the delightful but naïve pugilism appropriate to a
desert existence conditioned by independence, emotional intensity and
scarcity. There is no better way for Somalia to 'capture' its mobile
'peasantry' than by transforming them into the guerrilla combatants
that their mode of life predisposes them to become. It is a land approaching the size of Texas, mostly desert and scrub and parched pastures,
stretched along and down a tropical coast for hundreds and hundreds
of miles, without trains, with impossible to non-existent roads, with
virtually no industry, with little settled agriculture, and precious little
trade, occupied by a few million scattered nomads, topped by an
impoverished government. War, in such a setting, can be a boon – if not
to the people, then to those who govern them. Successive Somali
governments, civilian and military, have sought popularity and success
through belligerence. It is a policy, given Somalia's strategic emplacement, which has consistently, and literally, paid off.

The Somali government could not possibly pay for its artillery, fighter
aircraft, tanks and arms, or, on its own, train and equip every seventy-
third Somali citizen, as it did in 1985, to form an up-to-date army in the
order of 70,000. The military gobble up 65 per cent of the total budget.
They consume a large slice of all export earnings. The Somalis, obviously, do not themselves pay for these things. They secure arms, in the
present as in the past, because the industrial powers that manufacture
them have chosen, for their own strategic and other reasons, to serve
(without monetary recompense) as sources of supply. In 1985, keen to
purchase influence along the Red Sea corridor, the USA touched
Somalia with the magic wand of Yankee largesse, making the Somali
Democratic Republic the third largest recipient of US aid in Africa,
after (significantly) Egypt and Sudan. Somalia is far from passive or
selective in prosecuting this passion for military manna. Its adoption of
a kinship, which is in effect a 'racial', mythology, forearms it: whence
Somalia's dual sense of being wronged, and its justification for doing
wrong.

A Western donors' conference was called in Paris in 1983 which ended
by pledging to Somalia a massive sum, amounting to an average of US
$800m. for each of the years 1984–5 and 1985–6 – a sum equivalent to

the entire Somali budget in the year pledged. United States aid in 1985 alone was worth $100m. to Somalia, at least $40m. of this unambiguously earmarked for military purposes. Somalia's President Siad Barre has turned in every imaginable direction for more, stooped to bland threats, maligned his enemies, entreated those he would have as friends. Both Barre and Libya's Muammar Gaddafi came to power by a *coup* in 1969. Barre first turned to Gaddafi; in 1981 he broke off relations, partly to reassure the Americans; then in 1985 he healed the breach, largely to induce the West and the Saudis to take him more seriously, but also to hint to the Soviets at his availability and, following the account of Radio Mogadishu, 'to prepare to meet the danger and problems presented by subversive imperialist intrigue against the Arab nation'.

President Barre visited Saudi Arabia in 1985 to plead for millions of dollars worth of free oil, sufficient to cover the country's total annual consumption for 1986, and he succeeded – as in 1982, 1983, 1984 and 1985. He expelled the Soviets and Cubans in 1977, then assiduously and simultaneously courted the Americans. He received military aid from US allies – Egypt (Somalia campaigning in turn for Egyptian re-admission to the Arab League) and Saudi Arabia – and was blessed with a formal American agreement to supply military assistance in 1980. Displeased with US restraint in the business of restocking his depleted armouries, Barre attempted to renew his courtship of the Soviet Union – making heartfelt noises (no response from the Soviets) about 'normalizing' relations with a power consistently accused (to the West) of implementing an imperialist design against 'revolutionary' Somalia via the Addis Ababa government. The USA wants the man, of course, but is embarrassed by him; the irredentist war, after all, is supposedly a thing of the past. Like the Soviets, the USA attempts to give only as much as will keep the line taut, not to hold it so firm as to make it break.

Other means (which do not directly involve the USA) of achieving the same end of re-supply were found. The Somali Foreign Minister visited Pretoria in May 1984. Pik Botha, the South African Foreign Minister, was received in Mogadishu in December. The basic subject discussed presumably related to securing overflight and landing rights for South Africa in exchange for war *matériel* and military training for the Somali forces. The London *Observer* claimed (and the Somalis routinely denied) that the South Africans had already flown into Somalia plane-loads of equipment at night over the preceding month and a half, together with technical personnel. On and after 22 May 1985, major newspapers worldwide carried reports of claims by the Somali opposition movement that the Israelis were, by night, like the South Africans, flying into the Mogadishu area masses of military *matériel*, following a February 1985 visit to Jerusalem by a Somali delegation (led by the President's son-in-law, a general, and by the commander of the Militia, Brigadier-General Abdi Hussein).

It would seem fair to say that the Somali state, in the way that it has evolved, and without other significant resources save location, is in essentials a war machine. When the Somali government engages in war (as in 1977–8), it is also, whether voluntarily or not, engaged in the business of manufacturing refugees. The United Nations High Commissioner for Refugees (UNHCR) gave 700,000 as the figure for the Ethiopian (mainly Ogaden) outflow into Somalia after 1977–8, people now distributed over thirty-six camps in Somalia's south and northwest. If the new inflows reported for 1985 bring the refugee total in Somalia to 900,000, then almost one in six of all persons on Somali soil is to be accounted a refugee.

The actual numbers may well be lower. The Somali government figure for its refugee population is 1·2m. The UNHCR, which does not accept the government figure, has agreed the lower total, for funding purposes, of 700,000. But there are suggestions that a more accurate figure would be over 40 per cent lower still: around 400,000. There are of course incentives to any impoverished government to inflate refugee figures; Somalia receives aid for refugees that amounts, at a conservative estimate, to a fourth of its GNP.

Certain developments may, however, prove indicative. Arthur De-Fehr, the UN High Commissioner for Refugees in Somalia, was instructed from Geneva to clear up rumours about the inflation of numbers by Somali officialdom in order to increase income. DeFehr, by all accounts an inoffensive enough Canadian, was expelled from Somalia in June 1983, together with the UNHCR legal officer (a Ghanaian) and the UNHCR education counsellor (a New Zealander). These officials had reported behaviour consistent with the following pattern: a substantial Somali jacking-up of refugee figures; the induction of refugees into the Somali military; and the re-direction of funds for education from genuine refugees to finance the education abroad of the offspring or designates of Somali civil servants.

The picture of refugee resistance to repatriation is not altogether convincing. In *Les Réfugiés Africains* (June 1984), Dr Ibrahim Dagash, a Sudanese journalist working with the O A U, wrote as follows: 'Finally, a significant fact should be noted relating to our visit to Qoryoli camp [120 km. south-west of Mogadishu]. A group of family heads, Oromo people from Ethiopia, came together spontaneously before our group of journalists to say that they wished to be repatriated as quickly as possible, even if without assistance.' There are of course disputes between various, especially nomadic, groupings when brought together in the insufferable conditions of camps. But there is often enough, too, discontent with the way these camps are run.

For a start, it seems clear that food relief has not got through to all refugees actually in the camps: the fact that scurvy has broken out would alone suggest as much. Health conditions are more parlous than they should be, and some important part of this is presumably to be explained by official inattention, at the least. Reports indicate that, in the last week of March 1985, in the one northern camp of Gannet, near Hargeisa, cholera killed over 400 refugees. By 9 April, the deaths claimed exceeded 1,500. Further outbreaks were reported at four more camps. By 16 April, at least 3,000 cases of cholera had been reported.

The U N H C R claims that about 80 per cent of the refugees in Somalia are women and children. This is a high figure, but it is not inconsistent with what we know of camps in Ethiopia, Sudan and Djibouti. The younger men may not wish to come in – whether because of the boring, endless aspect of it all; or because of unhealthy and uncomfortable conditions; or because of a fear of victimization, one aspect of which is military induction; or because they have already volunteered to fight, or been conscripted, or relocated. A very high proportion of the young men missing from Somali camps are taking their chances in the bush, mainly on the Ethiopian side of the frontier, harassing (or being harassed by) the Ethiopian conscript levies opposite. This refugee population may appear to Somali officialdom to be easier to control than the normal nomad. These new camps, unknown in 1960, constitute a variant urban population, down on its luck it is true, but grouped together none the less; and the people can be taught, directed, perhaps transformed into equally controllable fishermen or farmers. Such, at all events is the hope, however bleak and forlorn, which tempted Japanese volunteers, in 1983 for the first time, at the invitation of the Barre government, to descend upon a virtual desert near the Juba River, to try to transform it and the refugees collected there into a more stable reality.

The people in the Somali camps are basically there because of a major war which the Somali government started. And this government, by now, has very probably lost all control. The 1977–8 war could not have been fought had the modern weaponry not been so indulgently supplied. The Americans (after the Second World War) first armed the Ethiopians; then the Soviets (after 1963) armed the Somalis; then the Somalis (1964 and 1977) attacked the Ethiopians; during which the Soviets re-armed the Ethiopians; after which the Americans and their allies re-armed the Somalis. Somalia has found it objectively difficult to repudiate this past, since only war seems to give it hope for a future. Somalia's refugee population is a consequence of its militarization, and also a means of enhancing it.

The Somali peoples are variably referred to as an ethnic group, a tribe, a nation. In all this there is intended a distinction between the Somali peoples, taken as a whole, and the state of Somalia, conceived as a more limited bureaucratic reality. It is difficult, within certain limits, to detect the aspirations of the Somali people. A chief aspiration of the state of Somalia, however, is plain to see: to make the Somali Democratic Republic co-extensive with the Somali people. The Somali people were never embraced by a single state system. But the imposition of colonial rule towards the end of the nineteenth century led them further in this direction than before. They fell under the control of the British, the French and the Italians. Following the close of the Second World War, these powers withdrew to be followed by France as much as seventeen years later (1977). Somali peoples today are incorporated within the new states of Kenya, Ethiopia and Djibouti, as well as Somalia. In the immediate post-war period (1946) the British Foreign Secretary proposed that all Somali peoples, together with the zones they inhabited, should be brought together under one administration, with a view to the latter eventually acceding to independent statehood. This was at a time when the United Kingdom controlled all but a tenth of Somali-inhabited territory. The British proposal failed to gain the assent of the super-powers and therefore was not implemented. But the aspiration preceded and survived the failure, first among Somali intellectuals, and latterly among the leadership of the Somali Democratic Republic.

The leadership of the Republic has regarded all Somali as kinsmen. Islam is sufficiently encompassing for it to have been adopted as the state religion. The assumption of common kinship and the fact of a common religion have tended to argue for some species of natural unity binding all Somali peoples together. Dr A. A. Shermarke, Prime Minister from 1960 to 1964, repeatedly argued, during his tenure, that all Somali peoples, inside or outside Somalia, were kinsmen, and that therefore the borders separating them from one another were not legitimate. For him, it was impossible to regard those who were genetically, culturally and linguistically related as 'foreigners'. (Dr Shermarke was ousted as Prime Minister in 1964 but elected by the National Assembly as the new President in 1967, thus displacing Dr Osman.) The government of Somalia in 1962 published a statement by Somalis in Kenya's North-West Frontier District which pointed in the same direction, towards a common kinship, a common brotherhood, within which context Somali political unity was demanded. President Osman, who assumed office in 1961, as first President of the Republic, also presented a similar case. He spoke (1962 and later) of the intolerance supposedly inevitable among a heterogeneous population, on the clear assumption that the Somali people, by contrast, were homogeneous – ethnically, culturally and economically.

On the whole, the Western press, but not African governments, seemed disposed to accept the Somali view. The position of the London *Observer* was fairly typical: 'Kenya Somalis ... differ from other Kenyans not just tribally but in almost every way. They are Hamitic ...' Naturally, Somalia was attacked from other African quarters on the grounds of 'tribalism', which was understood in this case to involve an attempt to detach from neighbouring states significant portions of their territory. The charge was answered before the Organization of African Unity in 1963 by President Osman, in a rather revealing fashion. The Somali people, he maintained, were not just 'an ordinary tribe without any rights to nationhood. The Somali people,' he insisted, 'are a nation in every sense of the word.' In spelling out what he meant by a 'nation', the President referred, among other things, to its distinct 'racial origin and characteristics' and to 'inborn qualities which render it indissoluble'.

The basic object of the Somali state, from its inception, was therefore the absorption of neighbouring territories inhabited by folk who shared a Somali identity. Such a programme is clearly a recipe for territorial expansionism. The Somali government, however, has always rejected the view that its policy is or ever was expansionist. Its primary concern, government spokesmen maintained, was to support the legitimate aspirations of other Somali peoples, not embraced by the Somali government, in their struggle for self-determination. If self-determination were allowed, the matter of union with Somalia could be entertained subsequently and separately, after the independence of the neighbouring territories had been achieved, and without prejudice to the continuing independence of the new territories, if they should choose to remain independent. Despite this line of argument the Somali government gave material support to secessionist forces in Ethiopia and Kenya. However conscious the Somali authorities may have been of what they were doing, the overall effect seems clear. To back *de facto* expansionism, whatever the adverse comment this generated abroad, had (domestically) a unifying effect. Conversely, to drop the expansionist programme made Somali governments highly vulnerable to internal opposition.

Somali nationalism, as distinct from West African nationalisms, has existed for many centuries. But it remains difficult to see what the object of this indigenous nationalism is. It is true that there are religious ties between the Somali. But similar ties obtain for example between Egypt and Libya, or between Switzerland and Germany, while the history of these countries has not and does not point in the direction of political mergers between them. It is true that there are genealogical ties between the Somali. But, again, genealogical ties, in themselves, in no way provide an adequate explanation of political unities. It is true that the Somali have come into conflict with foreign peoples, and this may have served to nurture a national consciousness. But it is also true that they have come into conflict with one another, repeatedly, providing the basis for many of the cleavages among them, and a basis for one of the commonest complaints to be encountered there – that against 'tribalism'. In one sense it is true that Somalia has no 'tribes', but only in the same sense in which the case can be made for all African states. For the very word 'tribe' creates or plasters over a whole series of problems and in itself explains little or nothing at all.

The Somali themselves do not hesitate to deplore the phenomenon sometimes covered by the word 'tribalism', nor do they hesitate to use the word itself. At the time of the army *coup* in 1969, the Supreme Revolutionary Council, led by Siad Barre, delivered itself, in Italian, of a brief but basic declaration of intent: '*Liquidare la corruzione, l'anarchia, il tribulismo ed ogni altro fenomeno di malcostume sociale nell'attività statale*' (my emphasis). The SRC declared war upon *il tribulismo*, among other social ills, understood as a form of kinship-based particularism. That declaration has been on display in banks and other public buildings of Mogadishu for years. Accordingly, we cannot take too seriously assertions about a prior national consciousness among the Somali.

Suspicion is appropriate in any case where this is asserted of any non-industrial people, and no doubt most especially where the latter are nomadic pastoralists. The political logic – and consequence – of subsistence economies is fission, not fusion: it consists of opposition to external, tax-imposing, centralized authority. Religious conformity and shared genealogies may paper over such individualistic, particularistic and centrifugal forces, but they cannot readily control or even contain them. Religion and kinship would more readily provide a basis for the expression of opposition to *prétendu* central authority. And such opposition will perhaps at times appropriately be labelled 'nationalism'. But to build upon it positively would provide a formidable challenge.

In a subsistence system, certain other related features follow: dispersal; the relative autonomy of different groups; the overlap between residential, working and kinship groups; the ideological, organizational importance of kinship ties; non-industrial production; the limited specialization of different roles; the tendency to define the boundaries of a group by reference to an overlap between territoriality and kinship; the recurrent tendency to resolve disputes with the centre through secession (direct physical withdrawal from control); the slow incidence of technological innovation; the vulnerability to bureaucratically and/or industrially organized societies.

The nineteenth-century European view of the so-called nation-state normally advances as its attributes the following: common racial identity, or even (which is to draw the line tighter) common kinship, common territory, common language and, in general, a common culture. In the African context, it is plain that no such formula, normally, will be held to apply. African states, which contain on the whole a great variety of conspicuously different peoples, would not qualify as states in the light of such criteria. Many African states, therefore, turn against such a model, and sing the praises of diversity. Many, by contrast – if not most – confess to such diversity, but only to stigmatize it as a form of adversity; and these sing the praises of unity, conceived as a goal to be won. But some of those who have a hope of living up to the European model, however dubious its value, may project (as a present reality) the existence for themselves of all those features contained within the European model. This, at least, has been the case with the government of the Somali Republic, since it first became independent in 1960, and indeed before.

It ought to be said that the earlier European definition of the nation-state grew out of a desire for regroupment, whether (as in the case of Germany) to create larger and stronger entities, or (as in the case of Eastern European and Slavic nationalism) to eliminate external control and create the basis for a more democratic society. Having said this, it is necessary to stress that an ethnically homogeneous society (Argentina under General Galtieri, for example) is not necessarily democratic and that an ethnically heterogeneous society (Switzerland, for example) is not necessarily despotic. One may understand a concern with ethnic homogeneity springing from a healthy interest in achieving representative government, and avoiding the reduction of class divisions to caste cleavages. But the concern with ethnic homogeneity also, obviously, can have a fascist or 'tribalist' dimension. Indeed, we might even conclude that any concern approaching an obsession with ethnic homogeneity may in fact be contradictory to a just system, meaning one that is genuinely representative, and in this sense democratic.

As regards Somalia it is plain that if we say that this state is optimally homogeneous, from an ethnic perspective; and if ethnic homogeneity is regarded as a sufficient condition for democratic government; then it is clear that the Somali case falsifies the thesis upon which projected Somali regroupment has impliedly been based. Somalia's ethnic composition has not changed since independence. But there can be no question that she is not today a democratic state in the sense hoped for at independence, and after.

Observers, domestic and foreign, have repeatedly drawn attention to what they have supposed to be the democratic thrust of the system, as derived from its relatively homogeneous base. I. M. Lewis, in the London *Times* of 16 September 1969, wrote as follows: 'The Somali republic is an unusual case in contemporary Africa. Unlike most other African States, which consist of patchworks of tribes and language-groups ... the Republic ... draws its strength from a long-standing ... sense of common ethnic identity.' Lewis concludes from this that 'The total effect is to provide conditions that ... allow a degree of internal freedom which the more fragile structures of other new African states cannot yet afford.' Within a month of this observation, which has been repeated in journalistic and academic quarters with the regularity of the Catholic Mass in Spain, Somalia's President Shermarke was shot dead by a policeman, and the government was taken over by the army immediately following the late President's funeral.

To understand the Somali system, what we must begin with is its basically subsistence nature. The regionalism, the clannism, and so many of the structural conflicts of the political system follow from this. Somali elections, from 1960 to 1969, clearly demonstrate the range of internal conflict. All were characterized by a growing profusion of parties and candidates, dominated by local and regional issues, largely to do, naturally, with bargaining over the allocation of very scarce resources. The internal competition produced the usual nepotism and graft, a process praised as democratic by most of those observing it at the time, at least up to the point of the 1969 *coup d'état*, when the virtues of the Somali political system were suddenly and radically reassessed. The civilian regimes which previously governed Somalia revealed, in short, the same dangerous faults as were to be observed elsewhere in Africa, whether in Ghana or Nigeria or Kenya, and which also led to unrest, to *coups* and/or attempted *coups* in these other cases.

Present-day Somalia was formed from a merger of two former colonial territories, Somalia (formerly Italian Somaliland) and British Somaliland. Separate elections were held in these two territories before independence. They reflect the pattern and scale of internal division which we would expect to be associated with the ecology. The first elections took place in Somalia (as distinct from British Somaliland) in 1954, followed by others in 1956, 1958 and 1959. Fifteen parties were involved in the first of these contests, each reflecting local interests and commitments. As in other African territories leading up to independence, the number of parties involved was significantly reduced by the time independence arrived. The Somali Youth League (SYL) was dominant among them, receiving most of its support from the Somali clans (Darod and Hawiye) which were coalitions of various lineages.

As for British Somaliland, the first elections were held in 1959 and 1960, just before independence. British Somaliland was smaller, poorer and even more (90 per cent) nomadic. Four parties contested the 1960 Somaliland elections, making a pitch for support to a population which at the time numbered only 650,000, most of whom, by virtue of being nomadic, were not genuinely accessible. The Somali National League (SNL) was dominant among these parties, receiving most of its support also from the Isaq clan. The British Protectorate became independent on 26 June 1960, followed by the Italian trustee territory on 1 July 1960, which also served as the earliest occasion for merger and the declaration of joint independence as the Somali Republic. The two legislatures merged, with an SYL leader from the South, A. A. Osman, becoming President of the Republic, another SYL leader from the South, A. A. Shermarke, becoming Prime Minister, while the SNL leader, Ibrahim Egal, was made Minister of Defence. (Four out of fourteen cabinet posts were awarded to SNL leaders.)

In Somalia's 1964 elections, the number of parties increased to twenty-one and the candidates to 973 – despite the fact that the dominant party, the SYL, won a clear majority of the seats (68 out of 123). By the time of the 1969 elections the number of registered parties further increased to about 130 and the candidates to well over 1,000 – with the SYL again winning (73 out of 123 seats). Twenty-six other parties won at least one or more seats. Many changes were produced by this last election: 70 per cent of all MPs (including SYL MPs) were ousted. Included among them were eight members of the government. (In 1964, by contrast, 55 per cent of all MPs were ousted.) In thirteen constituencies, no SYL candidates were returned at all. Indeed, only 40 per cent of the total vote was won by the SYL. During the elections, furthermore, twenty-five people were killed. There was ample evidence of votes being bought on a wide scale. In connection with this, seventy-seven petitions contested the results. When these came before the Supreme Court, the latter threw up its hands and forswore any jurisdiction in the matter. What is obvious from all this is that Somalis, at heart, entertained a great deal of suspicion and distrust for their elected rulers. At one and the same time, the electorate's wishes were narrow, selfish, regional (and in this served as a cause of the corruption of their leaders), while the actual behaviour of the electorate revealed a radical discontent which took on national proportions (and in this served as a cause of the intervention of the army).

After the 1969 elections the *Somali News* (4 April) bravely concluded that the results, the SYL victory, unmistakably showed that the bulk of the people put their faith in the ability of the SYL to lead them. John Drysdale, writing in the London *Times*, drew the same conclusion (16 September). C. P. Potholm in 1970 enthused generally over its 'impressive accomplishments in ... democratic decision-making'.[1] The

1. C. P. Potholm, *Four African Political Systems*, Englewood Cliffs, New Jersey, 1970.

evidence of such observers should be put against the large number of one-man 'tribal' parties still in existence in Somalia in 1969. It was obvious that the SYL had freely juggled constituencies to survive. Premier Egal was widely disliked, but he was none the less reinstalled. The 'opposition' (after the elections) refused to oppose, hoping to gain more – as elsewhere in Africa – by joining the government. They had done the same after the 1964 elections (making the government majority 88 out of 123). President Shermarke was himself a source of considerable dismay. This is not to suggest that the Somali people were prepared to rise as one man against their government. In a subsistence, and thus a markedly pluralistic, society, 'the' people are in an important sense not 'a' people – and thus *cannot* act so concertedly. What is clear is that few Somalis were happy with conditions in the country, and their own divisiveness helped to exclude the elaboration of a solution within the context of a conventionally 'democratic' parliamentary system.

Violence in Somali political life did not start with the army *coup* of 1969 and did not end there. Colonial states are violent. Colonial boundaries are arbitrary. The independent systems have inherited these features from colonial predecessors. They have also inherited the structural violence characteristic of these predecessors. Somali governments have said as much, and pressed for change. But they have pressed for change which could only be bought at the expense of their neighbours, and this pressure for change has itself produced violence – confrontation along the frontiers with Kenya and Ethiopia in particular. The pressure for the change of frontiers also conceals a domestic pressure for changes of a less obvious kind.

In the earliest phases of Somali independence there was an attempted *coup* (January 1962) by twelve northern officers which had as its object to detach ex-British Somaliland from ex-Italian Somaliland – to separate the north from the south. The *coup* collapsed within a day. The authors of it were arrested. All the same, this constituted a violent attempt at satisfying a grievance by resorting to secessionist means. The grievance of the *coup*-makers was that the north was less favoured than the south. And this particular grievance was of such importance that the effective punishment of its authors was not attempted. When war erupted with Ethiopia (1964) they were put back in uniform. In exchange for fighting against Ethiopia, they were pardoned for their earlier offence, which was technically nothing less than treason, in a special session of parliament. A clear pattern was set – the tendency to achieve internal peace at the cost of external adventurism.

In the early post-independence period, the thrust of violence in Somali political life became increasingly deflected into antagonism towards neighbouring states. In March 1963 several people were killed and seventeen injured when police opened fire on rioters in the northern town of Hargeisa. The people had been protesting against recent tax increases. In the Somali parliament in July 1963 about twenty-five opposition MPs attacked ministers, under-secretaries and members of the ruling SYL over objections to a draft law on public order which was felt to be unduly harsh and repressive. The police had to be called into the National Assembly to restore order. In December of the same year, the Governor of the Mudug Region was shot through the shoulder and critically wounded. This was near the regional capital of Gallacaio.

When the Somali Republic achieved independence in 1960, she laid claim to the French colony situated at and around Djibouti, to the Haud and Ogaden provinces of Ethiopia (about one-third of Ethiopian territory), and to the then Northern Frontier District of Kenya (about half of Kenya's entire area). Indeed, the Somali flag bears a five-pointed star, two of which points represent the merged ex-British and ex-Italian areas, the other three representing the non-Somali-controlled zones just mentioned. Even today Somalia has not renounced these claims. She has merely clarified the means by which she would seek to implement them, insisting that she seeks only self-determination for Somali populations in these zones. But this more sophisticated argument was only really advanced in 1967, at a point when Somali governments were finding the prosecution of war with their neighbours somewhat difficult to sustain.

At the start, the Somalis laid claim to the neighbouring areas of TFAI (Djibouti), Haud and Ogaden (Ethiopia) and Northern Frontier District (Kenya). The Somali government had sought to have the British government cede the Northern Frontier District (NFD) to the Somali Republic prior to Kenya's independence (December 1963). The British government set up a commission to investigate and it reported back that the majority of the people in the NFD would prefer to be linked with Somalia. The British government was also met, however, with opposition to any such cession by Kenyan representatives at the Lancaster House independence talks in London. Britain decided not to detach the NFD from Kenya. The Somali Republic broke off diplomatic relations:

she simultaneously lost a British subsidy of £1.5m. sterling. (And when relations were restored in 1967, this budgetary subsidy was not.)

There were border clashes and guerrilla raids and the most serious of these affected the Ethiopians. There was especially fierce fighting along the Ethiopian/Somali border in 1964, engagements which Ethiopia, U S armed, with superior equipment and larger forces, won. The Khartoum Agreement of the same year ended the fighting and led to the setting-up in 1965 of a Joint Border Commission between the two interested parties. These clashes, however, continued after 1964 into 1966, but on a much diminished scale, owing to a change of government in Somalia after the 1964 elections. In October 1964, at the UN, Ethiopia offered Somalia a treaty of friendship and mutual domestic non-intervention – which the Somalis rejected. At the OAU Conference held earlier that year in Cairo, Somalia opposed and refused to accept the resolution calling on African states to respect the frontiers inherited from the colonial powers. What became increasingly clear over time, certainly by 1966/7, was that the attempts so far made by the Somalis to extend the range of Mogadishu's authority, to form of the Somali people a single state unit, had failed. Accordingly, without renouncing the ideal of Somali ethnic unification, the newly appointed Somali Prime Minister, Mr Egal, negotiated an agreement in Tanzania with Kenya and Ethiopia to bring to a halt all hostilities. While Somalia sought to withdraw from direct military confrontation with her neighbours – a policy detected even as early as 1964, and accentuated in 1967 – she sought equally to build up both her economy and her armed forces.

In order to do this, Somalia had little choice but to fall back upon external powers for support. Typically, she sought support first from the British, the Italians, the West Germans and, most importantly, the Americans. But as the support received proved inadequate to Somalia's felt needs, she extended the range of her donors to include the Soviet Union, the People's Republic of China, Saudi Arabia and still others. One of the chief motives underlying Western aid was to exclude an Eastern presence, so that, when the realization of this object demonstrably failed, so did the logic of continuing most Western aid to Somalia.

There is no evidence to suggest that the *coup d'état* of October 1969 had anything whatever to do with Somalia's foreign policy, not at least as regards the continuation of the conflicts with her neighbours. Somalia had from the very start projected some form of expansion of her national territory. But no Somali government had clearly or fully thought through the means to be adopted to achieve this. The new republic, therefore, found herself at odds with all of her immediate neighbours. But they were all individually and collectively more powerful than she, and against them, in such circumstances, she could not possibly hope to prevail. The new military government of 1969 sought change, but no substantial change, in foreign policy.

The army was confronted with a situation widely regarded as corrupt, with elections plainly manipulated, and where there was little sense of the government being responsive to the electorate. After the 1969 elections the military were firmly determined to take over. The problem was more to do with timing than anything else. The assassination of President Shermarke provided more of a pretext than an actual reason for acting. The police were easily recruited to go along with the *coup*. And the reason given by army and police for it was 'the corruption of the ruling classes', characterized as having itself led to the death of President Shermarke. Charges were brought against the former civilian government for maladministration, injustice and, not least, tribalism, particular attention being directed to electoral irregularities, embezzlement of public funds, land deals, improper procurement of government cars and so on. The army gaoled the leading members of the government, including Mr Egal, the Prime Minister, and then dissolved the National Assembly, revoked the constitution and instituted rule by decree.

Despite the *coup*, the general trend of Somalia's foreign policy remained the same: the continuing concern with garnering foreign aid from whatever sources, for both developmental and military purposes. At independence, virtually all Somali foreign aid came from the West. In 1976, virtually all of it came from the East. In 1985, it was again all from the West, if we include Western client states, like Saudi Arabia. As Kenya approached full independence in 1963, it became clear that there was no hope that Somali demands for the (then) Kenyan NFD would be met. The Somali government, therefore, made every effort to overcome this setback by extending the range and complexity of its foreign involvements. In March 1963, a Somali parliamentary delegation, led by the President of the National Assembly, was despatched to the USSR for two weeks. In the same month, the Somali government closed its consulates-general in Nairobi and Aden, both still British colonies, in direct response to the British decision over the NFD. A Somali delegation went to China. In May 1963, China and Somalia concluded a trade agreement in Peking. In July of 1963, a five-man cultural delegation was sent to Peking, while an army general was sent on a brief tour of the USSR. Both China and the USSR began to respond positively to these overtures. In early December 1963, the Italian government ordered all thirty of its army and air force instructors to quit Somalia as soon as possible after 1 January 1964. This withdrawal was the first such consequence of Somalia's acceptance of Soviet military aid and her rejection of a joint counter-offer by the USA, West Germany and Italy. Western aid continued, but it was increasingly being overtaken by that from the East. Thus, in the period leading up to 1969, the USSR was estimated to have contributed up to 20·4 per cent of Somalia's development finance, with the USA in second place, covering 17·2 per cent.

The emergence of a military government in 1969 did not, then, coincide with any new departure in Somali foreign policy. American foreign aid to Somalia for the period 1958–69 is estimated at about $60m. This was cut off as of 3 June 1970, ostensibly because of trading relations with North Vietnam. But the more plausible explanation is that it was a reaction to the 7 May 1970 nationalizations of all major firms in the country (electricity, sugar, foreign banks, petroleum) taken together with the accelerating intimacy of relations with the USSR. The Somalis announced that West Germany had simultaneously cut off aid, because Somalia had recognized East Germany. (The West Germans, however, modified this by announcing that they were not cutting off aid, but would henceforth consider Somali requests for aid on a 'day-to-day' basis.) Within the week, a seven-man Somali delegation (including top officials from trade and industry) was despatched to North Korea, mainland China, the Soviet Union, Pakistan and Aden. In July 1974, a Soviet–Somali Friendship Treaty was signed by Presidents Podgorny and Barre in Mogadishu. The Soviets had previously acquired docking and other facilities along the Somali coast. Unpublished clauses of the 1974 treaty formalized and expanded these facilities. Somalia reportedly granted the Soviets full base rights and unrestricted access to all Somali airfields (which would permit the Soviets to undertake reconnaissance flights over the Indian Ocean). In return for these rights, the Soviets reportedly undertook to equip Somalia militarily for a period of ten years.

At the same time as strengthening her relations with the East, and with the USSR in particular, Somalia sought to integrate herself more fully into the Arab world. Such activity began well before the 1969 *coup*.

Somali representatives toured the Arab world seeking support for admission to the Arab League. Finally, on 14 February 1974, she was admitted to the League as the twentieth member. Somalia attempted, therefore, both before and after the 1969 *coup*, to secure sources of foreign assistance and support which would enable her to realize her long-term policy objectives, both domestic and foreign. She was very successful in this *vis-à-vis* the USSR and, in a lesser degree, *vis-à-vis* the Arab world, and subsequently *vis-à-vis* the US and European governments in general.

Somalia's long-term aim is to expand her territory and, it would appear, to tie down her errant populations. Virtually no country that today seeks to increase its territorial extent at the cost of neighbouring states can hope to do so by simple military attack. It will do so mainly by encouraging secession – on the basis of ethnic or other ideological affinity – followed by political assimilation of the seceded area or zone. In the period leading up and even into 1977, Somalia tried precisely this tack. But when it appeared that the prize might slip away, that the Soviets might re-arm Ethiopia before the WSLF and SALF could achieve their intended guerrilla aim, the Somali President could no longer restrain himself, and sent his own regulars on a vain and sanguinary last-ditch effort to secure the end of the rainbow. In this, the Somali concern was openly and unqualifiedly to wrest territory from her neighbour, whatever the justifications given for the affray. And it was the Somali side which was clearly in the wrong. The Somali military build-up, begun practically from independence and continuing today, is directly – and since 1977 demonstrably – linked to her territorial ambitions, to her kinship mythology, and to the cheap and destructive competition for advantage among her northern suppliers.

In confronting neighbouring regimes in the 1960s – most strikingly in the sharp military contest with Ethiopia in 1964 – Somalia (together with her guerrilla adjuncts) was always soundly trounced. The guerrilla forces whom the Somali government encouraged and armed were consistently overwhelmed by both Kenyan and Ethiopian regulars. There was considerable loss of life among the guerrillas sent across the border, their support bases were vulnerable, their equipment rudimentary. The arms build-up by President Barre in the 1970s was designed to overcome this deficiency. Barre sought to put pressure on the French in Djibouti, on Ethiopia and on Kenya. Indeed, French policy in Djibouti collapsed, followed by their withdrawal in June 1977, partly in response to the build-up in Somalia itself. But they left behind an elected government, protected by 4,000 French troops. In the circumstances, Djibouti ceased to be a credible object of Somali attack. Kenya, independent since 12 December 1963, had unwaveringly maintained its close ties to Britain, and increasingly to the USA, especially after 1969, which year effectively marked the climacteric of former Vice-President Oginga Odinga, a Soviet sympathizer, henceforth excluded from the counsels of the mighty. With Kenyatta at the helm, despite failing health, Kenya, in 1977 (especially since her 'NFD' had no significant population centres) was an even less attractive target than Djibouti. Ethiopia, however, was being bled grey by the bloody, slow-motion *coup* initiated in February 1974. She was confronted with a formidable attempt at Eritrean secession, complicated by four other subsidiary varieties of civil war, in circumstances (by 1977) where all US military supply had been shut down, leaving her acutely vulnerable to attack in her soft underbelly (the Haud and Ogaden).

Directly after independence, Somalia laid claim to vast surrounding areas on the assumption that relative ethnic homogeneity justified the assimilation of proximate areas under a single political authority. She pursued her foreign objectives fairly vociferously during the period 1963–6, albeit unsuccessfully. The Shermarke–Egal government instituted a detente policy from 1967 onwards. Somalia was quick to point out, however, that her objectives remained unchanged; what she did not want was to promote them in as directly violent a fashion as previously. When the new military government came to power in 1969, it was feared in some foreign quarters that a reversal of the new and more accommodating Shermarke–Egal policy might occur. It was a principal concern of the new Siad Barre government to allay these fears. The Somali Foreign Minister, for example, speaking in Nairobi (2 January 1970), reaffirmed that 'the new Somali government, under the Supreme Revolutionary Council, has no intention to deviate from the Kenya–Somali accord jointly signed at Arusha by President Kenyatta and the former Somali Government'. But neither this Arusha accord (1967) nor the earlier Khartoum Agreement with Ethiopia (1964) even stipulated any renunciation of long-standing Somali claims upon all surrounding territories. And the new military government gave no evidence of deviating from this apparent commitment either.

In November 1972, the Somali Ambassador to the UN (Mr Nur Elmi) revived old claims to what had been Kenya's NFD. The British government, by administrative action, eliminated this province, dividing it between Eastern and North-Eastern Provinces in early 1963, prior to Kenya's independence in December. But Somalia never recognized this act as legitimate, which explains subsequent reference by Somali officials to the 'NFD'. At the May 1973 OAU Conference, the Somalis again claimed part of Ethiopia as theirs. President Siad Barre (August 1973) characterized the control which Ethiopia exercised over the disputed Haud and Ogaden Provinces as colonial, and an obstacle to African unity. As for Djibouti, President Barre was even more forthright. Speaking in Ivory Coast (25 November 1974), the President described the territorial government (TFAI) of Mr Ali Aref in Djibouti as unrepresentative; termed as false President Giscard d'Estaing's asseveration that the people of TFAI wanted to remain French; characterized as an absurdity the 19 March 1967 referendum in TFAI, which he thought controlled by 'the guns of the Foreign Legion'; and stated, finally, that if TFAI were not granted its independence, the responsibility for what might happen in Djibouti would lie entirely with France. Ali Aref (an Afar) was subsequently removed from power, the pro-Somali government of Mr Hassan Gouled Aptidon (an Issa) succeeded him, with independence from the French following on 27 June 1977.

To achieve the objective of unifying the Somali people, President Barre required external support. To build up his forces with foreign aid, it was essential that his express concerns should be exclusively defensive. In the same way that we speak of garbage disposal as sanitary engineering, so do we now uniformly label ministries of war as ministries of defence. Among the various liberation movements established in Mogadishu after independence in 1960, the WSLF, then headed by General Wako Gutu, was destined to become the most important. His was not an especially successful tenure; the entire campaign, especially against Ethiopia and Kenya, had proved not only a failure but an embarrassment. It consistently frightened donors off. In the wake of the 1969 *coup* that brought Barre to power, Gutu, in 1970, was given short shrift by Mogadishu, and he ended up surrendering to Ethiopia. The Somali arms build-up, however, proceeded apace, now sponsored basically by the Soviets, subject to more centralized control from Mogadishu, and less subject to indulge in random attacks against indiscriminate targets, the only effect of which had been to make the rest of Africa exceedingly wary of Mogadishu. The W S L F was in fact clandestinely revivified, meshed more closely with the Somali army, and their operations redirected more especially towards Ethiopian targets, with activity in southern and eastern Ethiopia, from early 1976, being sharply stepped up. There were reports from Addis Ababa and from Mogadishu that several thousand W S L F guerrillas had moved into the Ogaden region of Ethiopia. Hassan Mahmud, now General-Secretary and Military Commander of the W S L F, claimed in late May of 1977, in Mogadishu, that his forces in the period from September 1976 to May 1977 had taken seven towns and destroyed six Ethiopian tank battalions involving something in the order of 3,000 soldiers.

During a meeting convened on 22 March 1977, in Taizz, North Yemen, between North Yemen, South Yemen, Sudan and Somalia, the President of South Yemen explained that the object of this meeting was to make the Red Sea into a region of peace and to ensure that 'neither imperialism nor Zionism' should control the area. Before the meeting began, the North Yemeni Foreign Minister, Abdallah al-Asnafe, in a press statement presumably directed towards Ethiopia, affirmed that 'the Red Sea is Arab'. The President of South Yemen was known to support some degree of accommodation between his country and Saudi Arabia with a view to promoting Saudi investment – despite the violence which these initiatives were ultimately to provoke within South Yemen in late June 1978. The March 1977 meeting was ostensibly held in order to accommodate the interests of states located along the Red Sea. None the less, Ethiopia was not invited to attend. Somalia did attend, being a member of the Arab League, although not linguistically Arabic, and despite the fact of not being located on the Red Sea at all (as opposed to the Gulf of Aden and the Indian Ocean). The North Yemeni Foreign Minister's declaration of intent – to turn the Red Sea into an 'Arab Lake' – had been advanced earlier and by others. But that sort of statement, in the circumstances, was understandably taken to signify that Ethiopia, from the perspective of neighbouring 'Arab' states, simply had no place on the shores either of the Red Sea or the Indian Ocean.

The conflict in the Ogaden had been building up all along, but it erupted in June and July of 1977 into a major war. One of the most significant of the opening phases was registered at the Kenya border post at Ramu, which, as we have described in Chapter 7, the Somali forces overran in making entry into Ethiopia. Kenya, in view of its weakened state and the complexity of the forces in play, was reluctant to become committed as a belligerent. But by September, with the dust settling, Kenya openly gave its backing to Ethiopia, despite her dependence on the USA and despite the significant difference in regime type between Nairobi and Addis Ababa. The Soviets, for their part, radically accelerated the delivery of arms to Ethiopia. Somalia, in November, more confident than before in her anticipation of vital military and financial support from the USA and Saudi Arabia respectively, resorted to the dramatic gesture of severing diplomatic relations with Cuba and ordering out all but a rump of the thousands of Soviet personnel in the country. Matters did not go Somalia's way, and the easy, early victories of 1977 turned into the bitter defeat and withdrawal of 1978. Somalia's Soviet-equipped legions were chased from the field by their newly equipped, Soviet-backed, Ethiopian counterparts. By March of 1978 it was all over as a war in the conventional mode. But, yet again, the Somali leadership would declare no peace. In January 1979, President Barre's Somali Revolutionary Socialist Party (SRSP), the sole legal party in the state, openly declared its support for the WSLF-led campaign, now reverting to its guerrilla form. Already in 1979, the WSLF and the SALF were again claiming control of the countryside in the contested areas: of course, on a barren, nomadic plain, there is little that cannot be accounted countryside.

The promoters of a Greater Somalia have made much of the natural ethnic ties obtaining between Somali peoples. In fact, the natural divisions that we may expect within any subsistence or semi-subsistence system, relating to population dispersal, heightened regional and ritual differences, a profusion of dialects and attendant loyalties, equally obtain among the Somali. The common kinship thesis underlying the Greater Somalia movement merely reduces to a most important mobilizational tactic for overcoming *de facto* social divisions which are at best inconvenient obstructions to a modern state which seeks to consolidate its hold on an economic basis that transcends the traditional (and decidedly weak) tributary form. There has been little evidence of unquestioning, automatic support for any Somali government, of whatever form, popularly elective or militarily autocratic. The military government of Siad Barre, which has had a life getting on for twice that of its civilian predecessors, has been no more successful in the achievement of Somali expansionist ambitions. Nor has the Barre government been exposed to any less opposition, it has simply been more ruthless in eliminating it.

Somali peoples will express their displeasure when their government extracts much and bestows little. There was a counter-*coup* against Barre in May of 1972, which failed, and was followed by the grim harvest (the July 1972 executions) that normally attends such reversals. Traditional religious leaders, foolish enough to assail in the mosques a decision to accord women full equality in matters of inheritance, were summarily executed (January 1975). Another attempted *coup* was essayed against Barre in April 1978, following the débâcle in Ogaden. The back of this upstart effort was also snapped, the usual executions following in October 1978. The attempted April *coup* was not officially read as an expression of public disaffection engendered by the costs of the government's own policies. It was denounced, on the contrary, as a mere expression of 'tribalism', a morally Neanderthalian excess, laid, on this occasion, at the feet of the Mijerteyn, a clan, it was held, burdened by an unfortunate animus against the clan of President Barre. It was mostly Mijerteyn who subsequently formed the Somali Salvation Front (SSF) which in turn evolved to become, by 1981, the Democratic Front for the Salvation of Somalia (DFSS). Another opposition group sprang into being in 1981, in this case the Somali National Movement (SNM), which Mogadishu promptly denounced as an expression of Isaq 'tribalism'. In February 1982, troops in northern (formerly British) Somalia dissolved in mutiny. In April, the austerity measures customarily imposed by the IMF upon ailing Third World economies sparked rioting. The SNM began to infiltrate northern Somalia with a view to taking out modest targets and mobilizing resistance to the Barre regime. In July 1982, the DFSS struck in the Muduq region of central Somalia. Naturally enough, the Ethiopians were blamed for these developments. And there can be no doubt that they had their role to play. The moral is no more than that of tit for tat.

The fact is that Somali ethnic homogeneity, and ensuing political stability, is more myth than reality, more aspiration than achievement. Somalia made an understandable constitutional mistake in attempting to build a polity around this mythology, legitimating the resort to force as a means. The life of peoples in the Horn has clearly been made more miserable in consequence. It is now being demonstrated that Somalia cannot with impunity direct guerrillas against her neighbours. In the end, they are perfectly capable of returning the favour. In the same way that Somalia backed the WSLF and the SALF and then sent their own units into action openly in support, so the Ethiopian government came in 1982 to back the SNM and the DFSS and doubtless sent numbers of their own men across the border to help secure the Somali frontier towns of Goldogob and Balenbale, while denying in the usual manner all responsibility. Dissatisfaction with the Barre regime, always helped along by the Ethiopians, is bound to swell. The difference is that the bent of the discontent is not the dismemberment of Somalia. For its part, the SNM, in January 1983, mounted a successful attack on the Somali prison at Mandera, facilitating the escape of hundreds of prisoners, both criminal and political. Reports continue regarding hundreds of casualties generated between government and rebel forces. Mogadishu's attacks, over state radio, on the Ethiopians are essentially racist both in substance and tenor. But all they really do is to reinforce the failure of Somali policy. The Ethiopian concern, as also that of Kenya, is not merely to secure a change of government, but a basic change of direction, away from the irredentist policies pursued since before independence. And that will be difficult. The fact is, however it may have come about, that the Soviets are now installed in Ethiopia, and that Somalia, willy-nilly, has become a client of the USA. The USA fell back upon the supposed Ethiopian invasion of Somalia in July 1982 in order to justify open, and no longer indirect, supply of arms to her client. The Pentagon announced that American forces would jointly participate in military ('Bright Star') manoeuvres, ending 11 August 1985, in Somalia itself. Washington now refers to Somalia as her closest ally in the Horn, which only means that Somalia under Barre will again be heavily armed and trained and that no denting of Somali irredentism is likely to be entertained.

The only real problem posed to the USA in its efforts to conciliate Somalia was and is Kenya, which is supposed equally to be a US ally, allowing free use of her port facilities at Mombasa. Just as the Soviets, by the interposition of Fidel Castro, sought to reconcile Somalia to Ethiopia in March of 1977, so did the USA, after 1978, become increasingly concerned to reconcile Somalia to Kenya. Kenya's President Moi, firmly set against Somali pretences, flew to Addis Ababa in January 1979 to make a public statement of his opposition to Somali policy. In the course of his stay, he signed a treaty of friendship and cooperation with the Ethiopian leader, Mengistu. Notwithstanding, American ties with Somalia grew firmer. In 1980, Somalia offered up her Soviet-built air and naval bases at Berbera to the USA for use by the Rapid Deployment Force. The RDF, later styled the US Central

Command, mounted joint military manoeuvres (in 1981 and subsequently) with various U S Muslim clients in the region: Egypt, Sudan, Oman – and Somalia. In return for the August 1980 concessions, the USA agreed military credits and other aid to Somalia. Substantial military aid was clandestinely delivered to Somalia by Egypt. Moi's visit to Addis was reciprocated by Mengistu in December of 1980, when the Ethiopian leader flew into Nairobi, and, with Kenya, jointly condemned the US pact with Somalia, promising the delivery of arms. But Kenya was in a bind. June 1977 taught her that she could not resist Somalia should the latter select her as a target for concerted attack. Nor could she expect any serious support from Ethiopia (and vice versa) unless perhaps Kenya could bring herself to embrace the Soviet bear. Kenya had been traditionally supplied by Britain, and this arrangement no longer appeared viable. She had no choice but to begin to shop for fighters and the like in the USA. And this was inevitably and firmly, more than before, to place herself in the hands of the USA. The USA apparently stepped up its diplomatic representations of concern to Somalia that the latter should at least mollify Kenya, if not Ethiopia. Indeed, it was only after the June 1981 OAU meeting, and only after Barre had entirely failed in his campaign to secure support in West Africa for Somali irredentism (now in the form of regional self-determination), that he was induced to see the wisdom of conferring with Moi – both sides under US prodding. The outcome was that Barre renounced territorial claims on Kenya, even going so far as to accept joint policing of their common borders. Thus the USA was successful in its move to conciliate its clients, where the Soviets failed. Barre, solidly backed by the USA, thus got his second wind – to further envenom the conflict with Ethiopia.

In case after case, the resort to arms in acutely impoverished states dries up development funds, redirects these to fuel the armaments industries of developed states and deploys the hardware obtained to demolish the fragile civil infrastructures that already exist. The great powers, slowly but surely, allow themselves to be drawn into these conflicts, and usually worsen them, by augmenting the flow of funding, by expanding the scale of death-dealing, thus intensifying animosities already sufficiently keenly felt. Somalia and Ethiopia and Sudan have all been flooded with refugees, outflow and inflow. They have been blighted by war and often wildly buffeted by economies blown irretrievably off course. Local leaders must assume their responsibilities. But then so too must the leaders of great powers who have taken consistent advantage of disadvantage.

# 10 Centre/Periphery:
## Ethiopian Imperatives

The bulk of eastern Africa, at the Horn, consists of a single state: Ethiopia. With an area of 472,000 square miles (1,220,000 sq. kms.) it is more than twice the size of France or three times the size of California. Two-thirds of Ethiopia form part of the East African Rift plateau and rise from five to ten thousand feet above sea level, high above the surrounding lowlands, both to east and west. The heart of Ethiopia is this high plateau, where the ancient kingdom of Axum was founded, well over two millennia ago, and where Menelik, in the 1880s, founded the present Ethiopian capital, Addis Ababa.

The sudden upthrust of the highlands has provided a defence against invasion from the north and west, from Arabia and Sudan. But to the south and east, the highlands achieve their elevation in an easy and less abrupt progression, and it is in these directions that Ethiopia has always been most vulnerable to attack, and also most open to the temptation of imperial expansion. These two are intimately related, after the manner of the faces of a single coin.

The Ethiopian plateau is itself larger than all of Spain; and it is comprised of three significantly distinct regions. (Perhaps it should be noted that the plateau is rugged, not flat, so that the expression 'plateau', from the start, is misleading.) The first part is by far the largest, and extends for almost the entire length of the country, from north to south, always lying roughly to the west of the Great Rift Valley. (It is only a part of this area – situated north of Addis Ababa – which is called the Amhara, or central, plateau.) The second part is much like the first, except that it occupies a smaller area, at a lower altitude, and lies always roughly east of the Rift. It extends up to the Ahmar and Mendebo mountain ranges, and ends more or less at the town of Dire Dawa. These two areas, despite minor variations between them, are normally fertile; in some areas the rainfall exceeds 100 inches per year; consequently, they are densely populated.

The third part of the Ethiopian plateau is the smallest. It lies south and east of the Ahmar and Mendebo mountains, south-east of the Rift.

Here, greens give way to browns, cattle to camels, and people to large, sometimes stunning, but empty vistas. The semi-desert, which constitutes a considerable part of this zone, is a preparation for a grandeur that lies beyond, yet lower and more barren still. This third part is called the Somali plateau. Through it lies the path to the eastern Ogaden, which speaks of deserts and nomads and Islam. It is, of course, the object of a great Somali ambition.

Ethiopia is almost exclusively an agricultural country. Its population, at about forty-two million, is the third largest in Africa (after Nigeria and Egypt). The rural population probably constitutes about 90 per cent of the total, and characteristically practises a combination of cultivation and pastoralism. The overwhelming majority of this population is located on what I have called the first and second parts of the plateau, with the greatest concentrations in the province of Shoa. Some areas in the centre of the plateau have population densities which exceed 400 persons per square mile.

Agriculture and animal husbandry account for 80 per cent of all Ethiopian workers. Service workers and professional people account for only about 2 per cent. Statistically, as in most other developing countries, agriculture is of declining importance *vis-à-vis* industry. Yet the *manufacturing* sector of industry still contributes only about 7 per cent of Ethiopia's gross domestic product (GDP), which is roughly double the proportion contributed by traditional handicrafts. Thus it is agriculture which is the economy's mainstay, generating 90 per cent of all export earnings, of which over 50 per cent is the gift of coffee. Virtually all of this trade has been and remains with the West. The cash crops for export, almost all of them products exclusively of the plateau – must find their way to the coast for shipment overseas. Hence the importance of Eritrea and its two ports, Massawa and Asab, on the Red Sea. And of Djibouti, in the Gulf of Aden.

Addis Ababa derives from the produce of its interior highlands. A coarse network of unpaved and potholed roads feeds these goods into the capital. (Asmara normally served a parallel purpose further north; at present it is almost entirely cut off by rebel activity in the surrounding countryside.) Along those roads is distributed a large and variegated population. The producers are overwhelmingly Amhara, Tigrean (as distinct from Tigray) or Oromo (Galla) peoples. There is rivalry between these three groups, but the future of the country, like its past, lies in their hands together. The element that is geographically most central and most cohesive is also that which has traditionally proved most clearly dominant – the Amhara. The Tigreans are both geographically and linguistically proximate to the Amhara; the Ethiopian state in its earliest form was founded by them. Together the two groups constitute roughly one-third of the entire population. The Amharic language is reckoned to be spoken by at least half the population; it is also the earliest vehicle of instruction in primary schools, up to the seventh year. The Oromo (Galla)-speaking peoples are more numerous than Amhara and Tigreans together but they are less cohesive than either, alone or both together. The Oromo, at 40 per cent of the population, are spread throughout the country, and have adapted differently, depending upon the setting. Whereas those in Shoa Province (about 4 per cent of the total population) are predominantly Christian cultivators, and intermarry with the Amhara–Tigreans, most elsewhere are not and do not.

One may see then that the highlanders cultivate the bulk of Ethiopia's produce and that they dominate the machinery of state. The centre of gravity of this control, over several hundred years, has shifted from north to south, but has basically always remained on the plateau. This is on the whole a settled, agrarian power and the nomadic pastoralists of the periphery, whether pagan, Islamic or otherwise, will likely remain alien to it, until such time as they too achieve the status – however unenviable – of a settled population.

Those who are both economically productive and politically dominant tend everywhere to allocate the bulk of what is best – in goods, services and jobs – to themselves. Thus the peoples who are peripheral to the plateau system are least well-off, enjoy the most exiguous benefits (schools, transport, hospitals, access), and in return are (on the whole) those least committed to the survival of the system. At the centre, there may well be bloody struggles over the political form the government is to assume; and there may be equally bloody struggles at the periphery. But it is more likely that the latter, whether in Eritrea, Ogaden or possibly (in future) on the south-western border with Sudan, will have less to do with altering the form of central government than simply with breaking away from its control.

The element, therefore, which imparts to Ethiopia such coherence as it enjoys is the geographically central, highland, agricultural, sedentary and Christian population of the plateau. It may be summarily referred to as Amhara–Tigrean–Shoan Galla. These peoples, taken together, come close to comprising two-fifths of the entire population. Despite internal differences, these groups have in some degree coalesced. Their coherence, which has evolved over hundreds of years, is the glue which has bound together the different parts of the Ethiopian state. It is not the case that other groups have no role. Some Gurage, Afar and others are also involved in the system, which, in its attempt to survive and expand, has described an outward motion of recruitment and absorption. And even the present revolution, begun in 1974, despite its bloodthirstiness, or even through it, may be seen as a confirmation of this.

The fortunes of Ethiopia, and of the centrally placed peoples who control her, have ebbed and flowed over time. The state was consolidated first in the north, and increasingly expanded to the south. Ethiopia has sought to defend herself against the sea, and yet to establish herself as a coastal power. She has sought continuously to expand and consolidate her highland base, and her very success has attracted hostile admirers. For many hundreds of years Ethiopia has been both in touch, and out of touch, with neighbouring powers, particularly to the north, east and west. The country has been invaded, pillaged and overrun by surrounding powers, her people well-bled, a target of forced conversion to Islam. Ethiopian rulers, in their own, austere self-righteousness, garbed in their own chequered cloak of Christian empire, have not been reluctant to return the favour. They have been repelled by a compulsory, statist and expansionist Islam. But they have not themselves been greatly disposed to the spirit of religious toleration: the Monophysite Ethiopian Orthodox Church has long since served as the institutional embodiment of an Ethiopian state religion.

The Amhara–Tigrean–Shoan Galla have fallen victim to European imperialism. Like the Hausa–Fulani in Nigeria, the Ashanti in Ghana and the Zulu in southern Africa, however, they themselves grew into a ruling element by imposing their imperial rule upon less powerful peoples. The Italians recruited the Eritreans and the Somali in the late nineteenth century in their attacks upon Ethiopia. The British in turn welcomed the help of the Ethiopians, at the beginning of this century, in their pursuit of the most famous of Somali national heroes, the Mullah. The forward process of Ethiopian expansion, alternating with the backward flow of invasion, has continued for centuries.

Imperial conquest, for whatever cause, is formally admitted today by no state to be legitimate. But all existing states, of any size, owe (in some significant degree) their origins to it. The evil inherent in it is unwelcome. The evil inherent in attempting, in the contemporary world, to undo it is feared by most states even more. Within Ethiopia's present territorial limits, given the country's diverse character, her subsistence base, her mosaic of customs and costumes and rituals and languages, together with her severely limited communications, there is unimagined scope for further coalescence. And this is also to say that there is the persistent fact of great tension, suspicion and bitterness, all growing out of the expansion – the imperial expansion – of a governmental centre from some small pinhead of power and authority in the distant past.

For a government to establish control over the coastal Horn alone would not greatly enrich it, apart from the economic advantage to be derived from the exploitation of the fertile region lying between the Juba and Shebelle rivers. Such a coastal government could not securely or profitably tax the nomadic pastoralists who inhabit the area. Either it must make the most of the strategic position of the Horn, in relation to the heavy commercial and military traffic passing through the Straits of Bab El Mandeb, or it must attempt to tax and otherwise profit from the trade of the highland interior with the external world – given that such trade must move across the coastal zone. It might of course attempt to do both.

Were one government able to lock up the coast entirely, through that it would control the external trade of the hinterland, and with such power as this it would certainly be able to exploit the strategic position of the Horn. But any great powers, which vitally depend upon access to the Red Sea via Bab El Mandeb, could not welcome such consolidation as this, unless they themselves instigated it – or, more pertinently, controlled it.

In the last century, the British and the French each warmly resisted the idea of the other acquiring total control over this crucial zone. The British attempted to work out on land the simple but compelling Cape-to-Cairo scheme worked up in the imperial map-makers' minds. The Germans cut across this ambition by occupying a part of what is today Tanzania, together with Rwanda and Burundi. The French acted similarly by supporting the continued independence of Menelik's Ethiopia, and also by initiating the construction of a railway across the British line of march.

The French, with the agreement of Menelik, hoped to establish a transcontinental railway under their control and/or influence, which would run from Djibouti, to Addis Ababa, to the confluence of the Sobat with the White Nile in Sudan, and roughly across the Central African Republic, Congo and/or Gabon (all French territories before independence). Such a link would have provided France with the possibility of a new route to the East, allowing her to bypass the British-controlled Cape of Good Hope. It would also have opened up the possibility of bypassing the only other existing route – through Suez. Finally, it would have enabled France to monitor, from Djibouti, all traffic in and out of the Red Sea through Bab El Mandeb. The British quashed such designs by the firm action taken in 1898 against the French at Fashoda on the Upper Nile.

It would be very difficult in ordinary circumstances for the coastal and lowland peoples of the Horn to resist Ethiopian peoples in their natural compulsion to move towards the sea. Of course, it would be possible for external powers, for strategic reasons, to ally themselves with coastal and lowland folk and attempt to restrain and divert the highland peoples further west. The natural tendency, however, from the time of the Portuguese (in the sixteenth century) forward, has been for any external power which sought a base in the region to ally itself with the inland peoples. This would appear to provide the firmest means of consolidating one's interests and authority at the coast. The Italians, in both the nineteenth and twentieth centuries, were not comfortable with the prospect of holding onto a lowland coastal area without attempting, also, to conquer the hinterland – on the assumption that only this would make the control of the coast secure. The French, hoping to settle at Djibouti in the nineteenth century, had no expectation that they could meaningfully continue there without some amicable arrangement with the Emperor and his successors. The Americans, following the Second World War, were quite happy to ally themselves with the Ethiopians and to cooperate with Haile Selassie in ensuring that Eritrea, after 1952 – and most dramatically after 1962 – was thoroughly and safely incorporated into the Ethiopian Empire.

What seems clear is that any government based on the coast, if its centre of gravity lies only there and not elsewhere, can scarcely hope to contain the hinterland, whose sedentary and agricultural forces seem destined, like the Nile itself, to burst through to the sea. Such a government, very much built upon sand – almost literally – whether or not it feels itself to be vulnerable, must, *vis-à-vis* the hinterland, surely prove to be so.

The reasons for this vulnerability are three. First, any government based on the highland centre, ringed in by forces at her coastal periphery which are united in opposition to it, will struggle assiduously to establish or maintain itself at the coast, and in every way possible to weaken the ties between its opponents. Second, no modern government can derive – as a government with a purely coastal base would have to do – a sure or extensive income from a population of nomadic pastoralists. Third, the highland government of the centre, however poor in absolute terms, controls an area which supports what is relatively a far larger population, a population which is more easily taxed and which yields a greater revenue from tax. Thus, in any sustained conflict between governments based, respectively, on the highland and coastal peoples, one would normally expect the former to prevail.

It is not at all difficult to perceive the greater power normally accruing to a settled agrarian state as against that available to a government layered out over nomadic pastoral elements. But it should equally be noted that a government of the second kind is weaker largely owing to the greater strength and independence of those individuals whom it has been set the task of governing. The structural weakness of this sort of government derives, paradoxically, from the relative invulnerability of its individual citizen.

A government based on nomadic pastoralism is structurally weak. Although it may behave harshly towards its citizenry, that is the least effective means of securing their support, since it is so easy for a nomadic subject literally to escape the attentions of his supposed governors. A government built upon nomadism cannot simply push the nomad aside – he is the only citizen it has. But it may, for a start, attempt to convert him into something else – such as a cultivator or fisherman. That is difficult to bring off, certainly to any significant extent. The government, alternatively, may simply leave matters as they are, hoping to survive over time, dreaming of an oil strike or some similar mineral miracle. This is also difficult, because it is from the young, with their energy and idealism, that the government draws its strength – and the young are not disposed merely to wait. The government may fire the nomad with its own city-bred enthusiasm for change, delineated in a manner which accords with the nomad's most natural inclinations. And one of these inclinations is to feud.

Where the highlander is verbally aggressive and argumentative, rendering insult an art and litigation a traditional way of life, the nomad is both hot and cold, and to insult him is as much as to stab oneself to death. To lead nomads is less to lead than to provoke, and such provocation attains its apogee in war. The nomad fights well, particularly as a guerrilla; his individualism is even enhanced by quarrels of that kind. But despite his energy and individualism, the future still lies with the agriculturalist, whether the nomad is beaten by him, or wins, and is absorbed by the imperatives of what he has defeated. The obligation upon the highlander, difficult as it is, is somehow to transform the pastoralist without destroying him.

The conflict between the highland, agricultural centre and the lowland, nomadic pastoral periphery has often been confused with a simple religious conflict, between Christianity and Islam. But it appears certain that the structural tension here would remain, even were the religious identity uniform, whether entirely Christian or Islamic or 'Pagan'. It is geographical difference, in the first place, which has favoured the establishment and retention of religious differences. The Christian peoples of Ethiopia – like those of Lebanon – were able to salvage their religious culture against Islam only through retreat to the magnificent, but rugged, terrain of the interior. Religious tension, therefore, although important, is not an entirely autonomous factor in explaining the continuing conflict between the Ethiopian centre and her coastal periphery.

Religious and ideological factors, whatever else we may say, can still achieve an independent emotional and political effect, directly influencing behaviour. So it is in the Horn of Africa. A great deal of the heat in the conflict between Ethiopia and Somalia has been fuelled by the perception which each is disposed to entertain of the other as somehow different, strange, perverse, wicked, perhaps heretical. The Ethiopians constantly remark upon what they describe as the unreasoning 'fanaticism' of the Somalis. The latter constantly remark upon the 'barbarism' and 'cruelty' of the Ethiopians. The observer is struck, when conversing with these attractive people, people who – even physically – have so much in common, that the one side is somehow deeply indisposed to concede the humanity of the other.

Ethiopia's identity has become, over time and at the centre, very largely Christian, but a population estimated at between one-fifth and two-fifths of the whole is Muslim. Eritreans are about equally divided between Christians and Muslims. Djibouti and Somalia are virtually entirely Islamic. Ethiopia is structurally the stronger power, and, if any change away from the hostility of the periphery is to be achieved, it goes without saying that a great deal must depend upon her. One strategy open to Ethiopia is simply to attempt to make her system of rule more egalitarian by incorporating more fully, on a power-sharing basis, the groups who are both politically and geographically furthest from the centre.

If a government which is identified with a specific power base – religious or any other – encourages the assimilation of elements who do not share this identity, it is also encouraging an outraged response from those upon whom its authority has most depended. It runs the risk of being overthrown by, or of becoming locked into a vicious civil war with, its own power base. Its internal opponents may argue that it seeks too much change – or even that it seeks too little, once the process of change has begun. Should either of these risks become a reality, there is the heightened danger of civil war at the periphery, where elements which wish to secede entirely from the centre seize upon a time of greatest disorder in the heartland to turn themselves into independent power centres.

Finally, there is the threat of neighbouring states, which see profit, or even justice, in the break-up of the dominant polity, throwing themselves into the *mêlée* to see if they cannot help to decide the outcome. If the dominant power is broken, the freedom of initiative of the smaller powers is enhanced. It is also possible that regions which have been wrested from the control of one centre may be assimilated – through physical absorption or play of influence – to the authority of those states whose arms or acquiescence helped to secure a secessionist outcome.

Ethiopia has for a long time been at risk, and so has it seen itself. It has attempted to expand, but it has always feared the expansionist designs of others. When its central religious identity was rather more crucial than it is today, threats were not only territorial, but also ideological. Because Ethiopia saw itself as a Christian country, based on conflict with coastal Muslim neighbours, it tended to turn in upon itself and to entertain a lively suspicion of the world around, and not merely in the end a Muslim world, as was demonstrated by the culmination of Ethiopia's sixteenth-century contacts with the Portuguese.

The Portuguese, in the sixteenth century, rescued the Ethiopian system in its Christian form. But although they saved the system physically, they also threatened it ideologically. Sixteenth-century Ethiopia's identity was religious. In Ethiopia, the Portuguese had sought out the fabled land of Prester John (as she was known in the medieval West) because her Christian identity established her as a likely African ally against the Turk. Portugal dispatched her seaborne soldiery, first to help, and then, in turn, to obtain help. Ethiopia was conceived as a central element in the Christian European scheme to outflank Islam, and thereby gain direct access to the riches of the East.

The searing conflict with both Islam and European Christianity in the sixteenth and seventeenth centuries gravely weakened Ethiopian institutions. The Galla or Oromo invasions which began in the same century in response to this weakness also further accentuated it, and these continued into the nineteenth century.

The traditional Ethiopian system was basically a tribute state. The Emperor was the recipient of agricultural goods and of military service, both supplied by governors or kings, who were either appointed or ruled by him. In such circumstances there is a great deal of regional autonomy. The problem with a system in which regional rulers enjoy considerable independence is that the system is strong only when the ruler is so, and threatens to fly apart when his fortunes fail; the centre is constantly at risk from the periphery.

Ethiopia has perhaps two million nomads and semi-nomads within her borders. Approximately a million and a quarter of these are Somalis (about 4 per cent of the population as a whole) located largely in the provinces of Hararghe (basically in the Ogaden zone of the eastern Somali plateau), Sidamo and Bale. These provinces run up to the borders of Somalia and Kenya respectively. The nomads overflow these borders. One of the regions they inhabit overlaps Ethiopia, Somalia and Kenya, extends for perhaps as much as 90,000 square miles (an area the size of the United Kingdom), has almost no rainfall and in the hot season is subjected to daytime temperatures which regularly range up to 100°F.

It is a zone that is decidedly peripheral to the cool highland centre. It is low-lying and almost waterless. It sustains population densities which may stretch to about thirteen per square mile, but are usually no more than one per square mile. This contrasts dramatically with densities in Shoa province of 400-plus per square mile. The region has little wealth; taxes (appropriated cattle) hurt its people more. Its inhabitants have few bureaucratic skills and less political voice, so that any injustice dealt out by those perceived as both callous and foreign, under the cross of an alien God, rankles the more deeply.

This talk of God cannot be lightly dismissed. There is very little left to the nomadic pastoralist apart from – as A. A. Shermarke declared in 1962 – 'the teaching of Islam and lyric poetry'. The Somali may readily fall back upon one or the other of these consolations, though it is normally difficult to divorce the two. The Somali conversion to a Sunni version of Islam was initiated in the ninth century from the seaports of Berbera and Zeila. The Somali have had considerable time in which to forge a sense, if not entirely of what they are, at least, of what they are not.

Too much emphasis can be placed upon their unorthodox habits, to suggest that adherence to Islam among the Somali is neither true nor consistent. There is heterogeneity, but the use of Islam as a religion, of poetry as a vehicle for expression (religious and political) and the assumption of a common pastoral culture all provide an important basis for the mobilization of certain forms of political engagement. The Somali are very different among themselves, and very divided, too, but there is enough that matters in common between them to permit them to act in unity against the forces and peoples of other religions and cultures.

Just as the observer may go too far in emphasizing the internal incoherence of the Somali on religious and other fronts, so may he go too far in emphasizing the internal coherence of the highland Ethiopians in parallel spheres. The degree of centralized government which it is possible for any subsistence or semi-subsistence system to achieve is severely limited. And the Ethiopian was and remains a form of subsistence system. Hunters and gatherers (like the Dorobo, Sanye, Boni and others of eastern Africa) are not known anywhere to have established large-scale systems. An obvious reason is that they are constantly on the move, must operate – to be successful – in relatively small groups, and have no means of rapid transport, for purposes of communication, whether by horse or camel. Nomadic pastoralists, even with camels and/or horses, are limited in similar ways, unless they can couple a psychology of harshness with rapid transport to turn themselves into a ruling element, nourished by the labours of more settled agrarian and commercial folk (the pattern, for example, of Arabic expansion under Islam). Thus it is basically only a population (*not* a ruling element) which is settled and agrarian that lends itself to the construction of a large-scale state system. This is what the Ethiopians achieved, and it is a goal which the Somali can scarcely be said even to have set themselves in advance of the present century. But even if the traditional Ethiopian system was integrated in a way unthought-of among the outlying Somali, it remained, for all that, a semi-subsistence arrangement, with characteristic limits on its potential for growth and integration.

Virtually the entire agricultural highland complex of Ethiopia is in normal times very productive. As we have seen, it will sustain a larger population than the drier lands to the east, produce a higher tax yield and permit (not necessarily create) a more elaborate form of government. There is thus a greater surplus of goods to be traded, though this is very small compared to that of virtually any industrial or semi-industrial state. In Ethiopia, statistically speaking, less than one person in ten lives in urban areas; in a country like England, less than one person in ten lives in rural areas; and in Australia, although it lives almost exclusively on the proceeds from an agricultural and extractive economy, the urban population is approximately 90 per cent. Production in the Ethiopian highlands is labour-intensive, is only elementarily mechanical and demands that much energy is expended solely with a view to supplying the labourer himself – the agriculturalist – with what he requires to consume. The volume of economic exchanges is limited.

The measure of this may be taken from the character of retail trade. Permanent markets, except for large cities like Addis Ababa, Asmara, Harar and Gondar, are not to be found. Throughout rural Ethiopia, which is most of Ethiopia, the goods of foreign manufacture are bought in regional markets; agricultural items of local manufacture in local, community markets; and neither variety of these markets will customarily convene for more than one day in seven (on different days in different centres).

The Ethiopian highlands, while constituting a centre in the ways we have already discussed, do not constitute a centre in any absolute sense. The plateau is a centre in relation to the periphery. But, taking account of its largely subsistence character, we may perceive cracks and fissures, glaring ones, too. Ethiopians do not trade a great deal with one another, or produce a great deal for one another. The landscape itself is rugged and broken, producing considerable isolation between different areas. In 1975 85 per cent of children of school age were not in schools. In 1985, the figure was probably nearer 50 per cent. Fully one-third of those in schools are concentrated within the capital province of Shoa. Perhaps half of all school children, who are taught in Amharic for the first six years, only learn it at school. Ethiopians, taken together, are reckoned to speak at least ninety-five distinct languages. In 1975, at least 90 per cent of the entire population was illiterate. In 1984, the UN gave a figure of about 92 per cent, and in 1985 would give no figure at all. Spread among these folk are greatly varying modes of dress, customs, rituals, observances and – not least – loyalties.

Thus, although one may observe that the Amhara constitute a quarter of the population, the Oromo 40 per cent, and that the former comprise a more coherent element than the latter, there are limits to such coherence. Even the Amhara – and however much they may have caused others to suffer – have themselves suffered much, and at their own hands, as it were. Among themselves, before the revolution in 1974, the Amhara were a highly stratified people. Even today, after so many years of turmoil, much of this remains. The Amhara were used to social interaction of an elaborate kind, to the forms of deference and accommodation which make stratified systems work. Perhaps half the population, up to the time of the reforms which followed 1974, were tenant farmers, owing and delivering a large share of their produce to their secular and ecclesiastical lords. Although the Ethiopian Orthodox Church probably owned very little directly, it probably enjoyed a right to revenue from perhaps as much as two-fifths of the land. The ordinary Amhara is not to be understood as having enjoyed significant advantages over less centrally located Ethiopian peoples, though his superiors may have done. He cannot aptly be portrayed either as living off the fat of the land, or as entirely lacking in resentment towards his rulers. The bitter civil strife, at its worst in Addis Ababa over 1977 and 1978, testifies to this.

Ethiopia has its centre, and although its centrality is relative rather than absolute, the economic activity of its peoples must seek an outlet to the sea. The economic lifeline of 'New Flower' – the meaning of the name which Menelik's Empress bestowed upon 'Addis Ababa' – is a stem of road and rail which reaches from the highland interior to the low and outlying coastal waters of the Gulf of Aden. The numerous and powerful highlanders feel naturally compelled, in as far as permitted, to force a path among the scattered pastoralists of the plains to reach the sea.

If Ethiopia were not located in the rugged terrain she occupies, she could hardly have held off European conquest for so long. But the people displayed an uncommon determination to hold out against European absorption. Ethiopia's survival for so long as a formally (if not substantively) independent entity reflected the strength of the culture which inspired this independence. The ability to recoup from Britain's invasion in 1868, and to contain the Italian variant in 1898, showed the resistance with which any attempt to transform the system – either from outside or inside – might be met. After the Second World War the very strength and tenacity of traditional forces, paradoxically, rendered Ethiopia one of the most conservative, indeed reactionary, states in the continent. What Ethiopia fought for, and mostly won – under Yohannes (1871–89) and Menelik (1889–1913) and Haile Selassie (1930–74) – was not freedom of a civil libertarian kind, but national independence of a kind best captured by the simple notion of legal sovereignty.

Thinking Ethiopians, despite the bitterness of protracted civil war (perhaps because of it), are now perfectly well aware that a simplistic traditional concern with the integral and unsullied legal sovereignty of the nation will not serve the needs of today. Perhaps the Eritrean war cannot be won; but the evidence, equally, is that it will not be lost. It is widely accepted, intellectually, that the simple idea of preserving Ethiopian sovereignty, however autocratic, is unworthy. At the same time there is widespread revulsion against the notion of the highland centre disintegrating.

The Ethiopians at the centre seem to be prepared to contemplate reversion to some sort of federal arrangement along the lines instituted *vis-à-vis* Eritrea in 1952 and revoked in 1962. It is precisely because it was revoked, and with such ease despite genuine opposition within Eritrea itself, that Eritreans are in no mind to think along these lines. They are perfectly right to rule out such a reversion: it would do nothing to protect their rights against what would continue as an otherwise monolithic centre.

The difficulty with the 1952 Eritrean–Ethiopian federal arrangement was much the same as that characterizing the 1972 accords (worked out in Addis under the aegis of the Emperor) between southern Sudan and the north of that country; and much the same again as in the 1964 arrangement between Zanzibar and Tanganyika to form Tanzania. In all cases, accommodation was sought between a smaller territory, either marked by dissidence or significant difference, and a larger territory, variegated enough within itself, but which confronted the smaller unit as a potentially overpowering partner.

This sort of imbalanced federation, of institutionalized amity between a territorial David and Goliath, need not be ruled out as unworkable. The bipartite structure of Tanzania shows that it is certainly capable of surviving. But it is less likely to work well and it will generate potentially damaging costs. The best we may expect from such a *de facto*, bipartite federation is that the larger of the two units will accord exaggerated place and power to the smaller in order to mollify it. This is exactly what has happened in Tanzania, where Zanzibar has been kept content by gross overrepresentation at the centre: Zanzibar, since union, has always been accorded the Vice-Presidency, and, in 1985, the Presidency itself. The risk is that the citizenry of the large unit will at the first hint of trouble unilaterally rescind such an unusual concession.

One cannot sanguinely fix upon the best that might happen with an imbalanced, bipartite federation. It is more important to contemplate the worst, which is that the larger of the two federated components will simply make a meal of the smaller. That is what happened in Ethiopia in 1962. That is also, in effect, what former President Gafaar Nimeiri caused to happen in Sudan from May 1983. Nimeiri changed the post-1972 balance of power with the South by unilaterally subdividing the region, with the effect and intent of weakening it *vis-à-vis* Khartoum. Just as the walkover in 1962 in Ethiopia immediately produced a continuing civil war, so did the Nimeiri move resuscitate the southern rebel movement.

On the whole, and whatever the difficulties, there is only one way in which most African states, especially the larger states, can move. This is in the direction of recognizing regional variation within the country, accepting it, surrendering the notion of physically extirpating it, surrendering earlier and wholly inapposite European models of homogeneous national mono-ethnic unity, according legal recognition to regional variation via some form of federal entrenchment, allowing particular regions to combine in the legislature with others that are like-minded so that no one region alone is ever likely to confront, in combination, the rest. Otherwise, African military establishments will simply continue to swell, and they will have as their job the business of bludgeoning a variety of sub-national clusters into submission.

But this is not just a problem for Africa; Africa is not on its own, nor has been for the past five hundred years. It is growing more involved with the rest of the world. And unfortunately, the more domestically representative any African state seeks to become, the more vulnerable does it appear to make itself to outside powers with global or regional ambitions and resources which seek, accordingly, to meddle. Africa is a continent of highly variegated populations which, under post-Renaissance market and strategic forces, are being pinioned to one another with a brutalizing speed. The problem is: how to make these mergers easier and, as an important part of this, fairer. This, indeed, is the challenge to Ethiopia: to engage in the tricky business of increasing the strength of the centre by expanding the representative force of the periphery. To do this will require outside help, not malice; it will demand of a future leadership, not the brute strength and paranoia of the Stone Age, but uncommon sensitivity, intelligence, restraint and courage. Nigeria, indeed, has been stumbling in this direction. What must be remembered, in observing these developments, is that what is involved is a new kind of state.

The Great Rift Valley runs south from the Dead Sea all the way to Mozambique. Addis Ababa sits on the western wall of the valley as it bisects Ethiopia. If one flies out of the city, south towards Nairobi, one is led to and past a succession of lakes before reaching Lake Turkana (formerly Rudolf) in Kenya. The Rift is a huge trench copiously filled with the waters of these and other lakes.

Addis Ababa, at over 7,000 feet, with a population exceeding a million, is the most populous eastern African city south of Cairo and north of Pretoria. The air is invigorating, the sun searing. Even a black skin, exposed for an hour or two, may burn and peel. Addis, given its elevation, is always cool by night, and by day (depending upon the season) skies are commonly overcast. The people are not unlike the weather. There is something inextricably sombre about them which suggests a burdensome history. Even in misery there is a sense of dignity, which is attractive, and a ready irascibility, which is not. There is nothing ancient about the city. It was founded by Menelik II (then King of Shoa) in 1880. Addis, as it has become popularly known, has none of the historical cachet of the ancient and more northerly towns like Gondar, Asmara or Massawa.

At the northern end of the city there is the lion house, with its dozens of captive beasts, perhaps intended by the Emperor to symbolize the perils of rule in so moody and precipitous a land. At the other end of the city, to the south, there is another interesting building. In the Emperor's day it might have been approached by descending King George Street, turning west onto Wingate Street, going as far at least as Wavell Street, then again turning south on Churchill Avenue, at the end of which one reached one's objective: the railway station. The lions come and go and appear to decline with every change of regime. Mussolini's Marshal Badoglio, who entered the capital on 5 May 1936 at the head of a motorized column of 1,725 vehicles, claimed that all the lions had been killed by the time of his entry. But the railway line has generally fared rather better. Among other things, it served Haile Selassie as a means of escaping into exile. All the streets around it he named out of gratitude to the British who restored him, with whatever reluctance, to his throne. Under the Dergue, these signs of honour were rescinded: the street names were changed. It would not greatly matter since the people of Addis do not appear to know the names of streets anyway, though they always know where they are. All of those English-titled streets terminated at a heroic French monument. At the front of this edifice, a sign reads, in large letters: CHEMIN DE FER. There it begins, as the name declares: a road of iron. This mean, metallic thing reduced the divide between the central highlands and the Gulf of Aden to a nod (one must still sleep on the way) and a rattle (the railbed, at under a metre across, is unmistakably narrow gauge).

It normally takes more than a full day to go from Addis to Djibouti by rail. On average, a total of thirty-two trains (twenty goods, twelve passenger) make the run each week, whether coming from or travelling to the Gulf. A passenger train leaving Addis on a Wednesday morning at 8.40 is expected to reach Dire Dawa on the evening of the same day. The Djibouti train carries both goods and passengers, but more of the former than the latter. The passenger service is the thinnest one might encounter almost anywhere. The Addis–Djibouti line is remarkably spindly to serve as Ethiopia's essential communications and trade link with the outside world.

When the line first came into use in 1918, there were no trains at night: active banditry did not allow it; travelling time between Addis and Djibouti was three days rather than a little over one. Contruction was long delayed because of sabotage by local peoples who lived along its path and because of Menelik's fear that it might supply Europeans with a high road to interference and eventually conquest. With sleeping cars added at Dire Dawa, the Wednesday traveller reaches Djibouti late on Thursday morning. The run from Dire Dawa up to the Ethiopian border town of Douenle is about 220km. Perhaps 85 per cent of the railway runs through Ethiopian territory; the remainder of the line, from the Djibouti border to the port, is only about 100km. (60 miles). (The French up to 1977 so arranged matters as to make this last sector the most secure.)

In the immediate vicinity of Addis Ababa, the land is normally green and fertile. Even, indeed, at Nazareth, perhaps 100km. from the capital and lower down, the land remains so. But as train and road proceed along the northern face of the Ahmar mountain range, up to and beyond Awash, the land becomes increasingly barren. Many of the people of Addis regularly retreat from their eyrie into the heat of the Rift Valley. When it is cool in the capital, they take the waters at Sodere – a small spa (which is very dry and almost hot) in the middle of the Rift, about 120km. from Addis, and just a little to the south of Nazareth. A Sunday afternoon draws quiet folk to walk and sit at the water's edge, to drink from or sink into it, sound and lame, all together.

Sodere is a tourist attraction, but its tourism is of a very indigenous kind. Cars are few. People seem indifferent to traffic. In this there is a remarkable difference between highland Kenya and highland Ethiopia. The tourism of highland Kenya appears to be almost exclusively reserved for aliens, mostly Europeans. In Ethiopia, it seems to be reserved almost exclusively for Ethiopians. The Ethiopian does not exclude the alien. But he is not so self-effacing as to deny himself the pleasures which his natural setting holds out. In a city like Naivasha, in Kenya, which has something of the same relationship to Nairobi as has Sodere to Addis Ababa, the African bourgeois is normally absent. Basically, perhaps the Kenyan feels *dépaysé* in a setting that remains predominantly European.

Travelling by train to Djibouti at the best of times can prove a zany and mildly hazardous adventure. Running time does not rigorously conform to any published schedule. No attempt is made to conceal the concern with security. A multitude of government soldiers filter into every nook and cranny, soldiers neither enthusiastic nor fearful, neither zealous nor watchful, however impressive their revolvers, rifles and submachine guns. Camels represent a more constant hazard than do guerrillas.

At Nazareth the line changes direction, from south-east to north-east, as does the new Addis to Assab road, and enters the Rift Valley. From Nazareth to Awash is about 155km., under 100 miles. About half way between the two begins the Awash National Park, which stretches on almost to the town of Awash itself. The hills here are broken, abrupt, hot, dry – volcanically upthrust against a clear sky. As the passenger lurches through this dramatic landscape, he is confronted with a remarkable sight. A volcanic cone, so large that it might pass as a combination of several, looms large to the north. Its size is such as to drown out the visual effect of everything else in the landscape. This is Fantale, grand but extinct, a volcano whose 15-km. diameter accords well with the passenger's sense of its extent. Beyond, roughly from the town of Awash, the Rift Valley begins to lose its elevation; it descends into the Danakil Basin (in Wollo Province), and further north still into the Danakil Depression (of Tigray and Eritrea); there it is transmogrified into one of the fiercest deserts under the sun. The train continues beyond Awash through heat and dust, keeping just clear of the Danakil Basin for as long as it can, all the way into Dire Dawa.

There are goats everywhere, some cattle too, but none especially fine. Increasingly, as the train – locked between the Danakil to the north, and the Ahmar chain to the south – gradually and gently descends towards the coast, it is neither cow nor goat but the camel which comes into its element. This country is repeatedly cut by gorges, ravines, massive shifts of land. The camel usually moves through it all dependably, however unattractive in form. It proves its utility, despite man's low esteem for its intelligence. Frail bridges and fleeing camels, which together converge on an exiguous railway line, render life perilous. Coming round a descending bend, which leads down to a chit of a bridge, high-perched above a dried-out stream (or even, depending upon the season, a river in spate), the train may be shaken and its passengers startled by a dull thud. Such sound might signal an overflow of war into an already shaky existence, but it is more likely merely to signify the demise of another camel and some further delay to the expected time of arrival in Djibouti.

Dire Dawa, an even newer city than Addis Ababa, was created by the railway. Like Nairobi, it served as the most important railhead for the line running inland from the coast. It was founded at about the time that the railway reached there from Djibouti, in 1902. It is inhabited for the most part by Amharic speakers and, in normal times, is full of camels and donkeys, which are brought to market here.

Beyond Dire Dawa and up to the Djibouti border, the railway branches off more sharply to the north (as does the road, for that matter). The 225-km. run up to the Djibouti border takes one up and down – but mostly down – into country which has an unmistakably *Beau Geste* quality about it. The hills are still rugged, but their elevation is more modest, and the air, though dry, permits the sun to weigh far more heavily. The passenger train normally goes through Douenle, the tiny Ethiopian frontier town, at night, whether from Addis or Djibouti, so that it is not easy for the traveller to inspect his surroundings. But in the broad light of day, the scene is dauntingly stark and forbidding. The tiny town takes on the character of a Stone Age encampment. Boulders litter the place and dwellings are constructed of them. One of these, before Djibouti became independent, housed the Mouvement pour la Libération de Djibouti. The MLD promoted the object which its name proclaims, but its membership was basically Afar and its finances supplied by the Ethiopians. It enjoyed a severely nominal presence in Douenle and was otherwise observed to be about as dead as the bodies consigned to the boulder-strewn graveyard which litters the north-east corner of the town.

The Djibouti frontier post at Ali Sabieh is more impressive than its Ethiopian counterpart. Ali Sabieh even has its own little fort, kept trim and white. Before independence, in 1977, the French were far more meticulous in protecting the railway in their sector than were the Ethiopians in theirs. The Somali government in Mogadishu had placed as its highest priority the expulsion of the French from Djibouti. It was not a challenge which the French took lightly, although ultimately they succumbed. The bridges within the former French territory, which are the railway's weakest links, were formidably well defended up to Djibouti's independence in 1977, after which there was not the same need, since those being defended against were largely in power.

Barbed wire was massed on the approaches to the bridges above, and spread like fields of sandspurs in the gorges below. Towers with powerful beacons were installed at both approaches, and helicopter landing pads were ready nearby. Machine guns were mounted above the coils of wire. Presumably, at the perimeter of the gorges below, minefields had been laid. It would be almost impossible to break through such defences as these without at least signalling one's presence. And if a presence were declared, it only required a moment for reinforcements to be flown up from camps nearby, and ultimately from the city of Djibouti itself. The French made extensive use of armoured cars and personnel carriers further along the railway, beyond Ali Sabieh. Virtually the entire area that fell under their jurisdiction was broken desert or semi-desert. Mirage jets flew protective cover overhead. The trains themselves, on entering Djibouti, were thoroughly searched by Foreign Legionnaires in shorts, khaki and kepi, wearing sidearms in black leather holsters, many of them bearing submachine guns. The situation on the Ethiopian side of the border appeared much more relaxed. But beneath these appearances, the burgeoning reality was not such as could offer any objective consolation to the Ethiopian authorities.

The railway, running through the semi-desert and desert zone north-east of Awash, is particularly vulnerable. The rugged terrain often twists and curves in a severely unaccommodating manner. Virtually all the area that lies beyond Dire Dawa and towards Djibouti is inhabited by Somali-speaking peoples. In this region there are no significant natural or geographical or human differences which correlate with the political gulf that separates Somalia from Ethiopia. The Ahmar Mountains spiral across into the North-West Province of Somalia. The Somali nomads round about readily ape the movements of these mountains. North-West Somalia has no geographic identity which permits it to be distinguished from south-east Ethiopia. If a border is drawn at all, it cannot but fail to be drawn arbitrarily, taking account of the human populations which will be spread over or obstructed by it.

Ethiopia's terrain, particularly on the central plateau, provides defence against external aggressors. By the same token it encourages internal dissension – indeed rebellion – by making it so easy for dissident elements to hold out against the imperious demands of any central government. Ethiopia, even under the Emperor, had serious difficulties with Eritrea. Dissident elements were not only to be found in Eritrea, but throughout a solid wedge of other northern provinces (Begender, Tigray, Wollo and Gojjam) bordering on Eritrea, all covering approximately one-third of Ethiopia's territory. The Tigray Secessionist Movement, the Tigray Democratic Movement and the exiled Ethiopian Democratic Union, all operating in these areas abutting on Eritrea, were, like the latter, equally up in arms against the central government. The heightened disaffection following 1974, not only in Eritrea, but also in adjoining provinces, signally contributed to the turmoil which embalmed the Dergue.

It was even less likely that all of these provinces should unite than that they should suddenly concede uninhibited support to a new military government in Addis. Since 1974–5, the northern third of Ethiopia has been open to physical attack upon civilian vehicles, military convoys, road and rail, police posts and villages; there have been kidnappings and the like, so that traffic cannot securely circulate unless – and sometimes not even then – under police or military protection. Hence the importance of routes to the sea from the capital which bypass the northern centres of population.

Even at the most placid of times, the idea of the Ethiopian centre and southern provinces conveying their trade to the sea via the forbidding marches of the northern highlands would prove a grossly uneconomic proposition. It is a long and arduous journey from so southerly a point as Addis Ababa to so northerly a port as Massawa. The land through which road and rail must run is staggeringly difficult. The idea of building a railway across such terrain is nightmarish. But access, actual and potential, will naturally be insisted upon. It cannot be Addis Ababa's first and rational choice to reach the sea via the long and rugged route to Massawa. The easier and more rational way would be to beat a path to Awash and then to strike north through the Rift for the port of Assab. (In fact, the road from Awash to Assab, stretching for over 700km. was opened at roughly the same time that Haile Selassie's fifty-eight-year reign was brought to an end by the Dergue.) The alternative: to continue north-east from Awash along the northern face of the Ahmar mountain chain to Dire Dawa, and at that point to strike north for the Gulf of Aden at Djibouti. In either case, whether the choice is to move to Assab or to Djibouti, it must follow the declination of the Rift Valley as it moves down to the Red Sea and the Gulf of Aden. The railway to Djibouti was originally conceived as the most important outlet for Ethiopian trade. The build-up of Assab is a more recent phenomenon. The Djibouti railway proved difficult enough to complete, but doing so represented a far easier option than attempting to construct a line running from Addis to Massawa.

The ideal for Addis Ababa is to secure an exit to the sea at a point that is relatively secure, from a military point of view. An exit to the sea at Assab has the drawback that it lies just above the narrow Straits of Bab El Mandeb. Traffic which is caught to the north of that choke-point is clearly more vulnerable than traffic which enters the Gulf further south. Djibouti gives access to open sea and thus constitutes a safer port. The Ethiopian centre, whatever may happen to the Eritrean north, will always wish to maintain a coastal outlet which skirts the northern reaches of the Rift Valley and which by preference issues onto the Gulf of Aden rather than onto the Red Sea. At the same time, given the trouble in the region, Ethiopia could never freely contemplate surrender of either of the Eritrean ports it now holds: to lose Assab or Massawa could invite anew the closure of Djibouti.

The Ethiopians insist upon sovereign control of at least some of their coastal outlets. Their detractors argue that there are many landlocked states and that there is no reason why Ethiopia should not be incarcerated along with the rest of them. Ethiopia, however, is in no way a minor African state. It is the third largest country by population in the continent. It should be one of the most agriculturally fertile, given peace and sensible government. Nigeria, with a population over twice as large as Ethiopia's (and an area 25 per cent smaller), has immediate access to the Gulf of Guinea. Egypt, with a population 25 per cent larger and an area 20 per cent smaller, is open on two of her four sides to the Mediterranean and Red Seas respectively. The Sudan has direct access to the sea through Port Sudan. Zaire has a tiny foothold on the Atlantic via Matadi and Boma (due to the manoeuvring of colonial antecedents).

There are geographically large countries in the continent of Africa (Mali, Niger and Chad are all bigger than Ethiopia) which are denied access to the sea via ports subject to their own sovereign control. But there is virtually no country of the order of importance of Ethiopia which finds itself in such a position.

# 12 Periphery/Centre: Djiboutian Derivatives

In the nineteenth century, France sought to trade with the interior, with the peoples of the Ethiopian plateau. Just as the British pushed inland from the Swahili coast towards the fertile lands of the Kabaka, in Uganda, so the French moved inland from the Somali coast to the fertile lands of the Negus, in Ethiopia. In both cases, a caravan trade was initially established. But whereas Kenya's Mombasa had existed for centuries, the city of Djibouti, which was never incorporated into Ethiopia, was created almost overnight. There had been historic settlements further south at Zeila, for example in what is now northern Somalia, but nothing whatever at Djibouti.

Before independence, and indeed for more than two years after, virtually nothing was produced in Djibouti itself (apart from a few soft drinks). Djibouti's water is saline. It will not nourish vegetable gardens. The celebrated *gat* (spelled in so many different ways) that Djiboutians chew, the *doura* they down, their tea and coffee and sugar and milk and flour and rice and meat (when they have it), and the *injera* (a species of bread) that some adore – whether made from the Ethiopian *teff* or from Somali sorghum – all come from elsewhere.

The ocean around may abound in fish, but Djiboutians, like most nomadic pastoralists, prefer to leave them there. The Afar have no taboos against fish-eating. But it is not an Afar town. Folk do fish, but fishing is an engagement normally left to the Yemenis, who hail from a land north of the Gulf, on the other side of the Red Sea. Fish belong to the sea, Djiboutians to the land.

Indeed, the President of the new republic quite simply refers to his compatriots as 'a people of herds and deserts'. But the new Djibouti which Mr Gouled leads exists more in spite of, than because of, herds and deserts. It exists because of a port and a railway. Two-thirds of Djibouti's total population of about 330,000 is concentrated in the city of Djibouti itself. They survive, for better or worse, through the limited employment the city is able to offer, deriving from its function as a commercial waystation or entrepôt.

The population, whether it speaks Afar, Issa, Arabic, French or Amharic (or even English), is almost entirely (92 per cent) Muslim, setting aside the diplomatic corps, migrant highlanders and of course the French. (The French presence is pervasive, whether administrative, commercial or military.) Despite this Islamic character, it is not so much the *muezzin* calling the faithful to worship whom one hears. It is a bugle sounding reveille, at 0600 hours. It is an African crow cawing, a donkey braying. The ulema is more muted and heard only faintly, if at all, as if to suggest that the message of the Prophet ill accords with the relaxed, even abandoned, commercialism of the city.

For the poor, which means for almost everyone, life in Djibouti is hard. At independence, in June 1977, observers estimated that, out of the entire population, no more than 18,000 had jobs, say between 7 and 8 per cent of the total. Djibouti officials, at the same time, gave an unemployment figure of 80 per cent. The UN gave a figure of 50 per cent for 1984. It is best to treat these figures with a degree of scepticism. But the observer's intuitive impression in 1978 was of a very high unemployment rate, and in 1985 of a rate still high, but reduced. Despite the presence of 16,000 refugees, Djibouti's 1985 inflation rate was reduced practically to zero, and a substantially successful effort has been made to contain the pugilism of neighbours at the frontiers.

Virtually all of the buildings of the city centre, perhaps a mile square, are wreathed in colonnaded arches, which protect the passer-by from the direct glare of the sun and from the very occasional drizzle. The soil of the city seems uniformly clayey, and even when wetted, normally rejects the relief it is offered. The centre of town, socially if not spatially, has about it a comfortable air, a casual effervescence. The square is ringed by shops and bars and hotels, nearly all run by Frenchmen and women, with a smattering of Syrians, Greeks, Yemenis and others, all catering to the transient trade of the port. The taxis, all neatly aligned and regularly washed, driven by Afars, Issas and even Ethiopians, are large and clean and inviting compared with those of Mogadishu or Nairobi; they are also cheaper to hire – one of the few bargains which Djibouti has to offer.

There is the formal trade of the shops, each ringed in by high white walls and large glass windows and blessed with a halo of ceiling fans, eternally revolving (their wiring is often precarious). In the odd record shop, or chemist, the delicious taste of air conditioning is welcome. But there is also an informal trade, beyond these neatly enclosed establishments, a trade conducted under the arches, but also beyond, almost spilling into the roadway itself – a trade in Ethiopian carpets and handicrafts, in heavy metal pendants and bracelets, in ostrich eggs and a variety of skins.

In the central square, in 1976, few skins were to be seen – perhaps two or three zebra. In February 1978, by contrast, there were dozens of zebra skins, a number of lion skins, and at the eastern end of the square as many as sixty skins of leopard and cheetah, mostly cheetah, and all in the custody of a single young entrepreneur. Very few if any of these animals were to be found anywhere near Djibouti. Where, I asked the dealer, did he procure these pelts? 'Western Somalia' was his answer. And when he said '*Somalie occidentale*,' there was conviction in his voice. We smiled in a conspiratorial understanding that what *he* meant by Western Somalia, the rest of us would take to be north-eastern Kenya and south-eastern Ethiopia.

In the late seventies and early eighties, poachers appear to have been busily at work throughout eastern Africa, certainly in Ethiopia, Sudan, Kenya and Uganda. They were dispersed over thousand of square miles, on foot, in jeeps, in Land Rovers, sometimes in light planes; very occasionally even helicopters were brought into play. They struck both into the Ethiopian and (more especially) into the richer Kenyan game areas. Uganda was a vast killing ground, not only for humans – in a sense, not even especially for humans. Hunters, in an uneven chase, were at work in the Danakil Reserve, the Awash National Park, south into Marsabit, Samburu, Meru. Many of the local poor were involved, but that was not new. As in Kenya and Sudan, many of these killers were close to, even in the pay of, powerful political figures, but there was nothing new in this either. What was demonstrated in Djibouti, after the 1977–8 Somali–Ethiopian war got into its stride, was that the slaughter of exotic animals by the common man had arrived. The qualification is that this common man was a soldier. In the Horn, with the outbreak and intensification of full-scale war, animal life was decimated, butchered for sport, as a distraction or for profit. The obvious outlet for very much of this cheap and grisly trade, throughout the Horn, was Djibouti. Cheetah skins in 1978 were selling in Djibouti, in the open air, for 18,000 Djibouti francs: $95. Zebra pelts were dearer, at 80,0000 francs: $421. The atmosphere was all Old Frontier. Buyers were not unduly hesitant, whether German, Yemeni, French or other. But they were mostly French soldiers: the pelt was like a medal; it signalled service.

War intensified the trade in animals, at least in their remains. As a matter of course, the trade could only flourish for a limited time, for the more successful it proved, the more readily would it subvert itself. The animals, at this rate, just could not possibly last. Kenya, over 1977 and 1978, had been persuaded to this view and simply moved to ban hunting and the attendant trade in ivory. It was not an easy move to make and was questioned on a variety of grounds, not the least of these bearing upon inefficacy. The professional hunters contended that the ban on their profession, since it removed them from the field, only made it easier for poachers. The problem for the government was how to justify to the population a measure that allowed rich foreigners to make sport with game animals while local folk were being entirely barred from traditional practice in the same line. Whether to good effect or not, the ban came down in Kenya. In Djibouti, there was nothing of the sort.

The Ogaden war had not created this abhorrent trade in animals, but it had intensified it to an unspeakable degree. The war had crippled the railway, and fairly pummelled the port to sleep, but simultaneously generated a trade in riddled animal hides, pegged out to dry, strung up on display. In February 1978, Djibouti's main square was piled high with skins. I clambered aboard a little nineteen-seater Djibouti airways DHC 6 to survey the sea and cross to Aden: Djibouti harbour was empty. I could spot below, off Le Heron on the Djibouti side, just one, solitary, stationary ship.

Within a twinkling, we were at Aden, with its competing harbour where a dozen ships leapt into view. For the first six months of 1976, the average number of ships calling at Djibouti was 143. For the first six months of 1977, the average was 100. In the first two months of 1978, there seemed to be virtually nothing at all. Such was Djibouti's great trade in time of war. Insecurity and war had cut off Djibouti from legitimate trade with her hinterland, and through no fault of Djibouti's.

In 1978, Djibouti had an air of being alive, even if in decay. The countryside, at a distance of hundreds of miles from the Djiboutian epicentre, was being savagely knocked about, its peoples uprooted and blown away, wildlife ceaselessly, perhaps irretrievably, destroyed, to nourish a deathly commercial liveliness concentrated in the few square feet of Menelik Square, so appropriately located at the Gate of Tears (which is a fair rendering of the Arabic 'Bab El Mandeb'). Men and animals were being put down with energy in Somalia, Ethiopia, Uganda, southern Sudan, north-eastern Kenya. Ivory, especially, was making its way to a variety of coastal outlets under the magnetic impulsion of money, which brought it commonly to Hong Kong, and indeed to China, where superb craftsmen might chisel it into the form of delicately poised dancers, or fashion from it burnished images of happy, potbellied Buddhas, whence it would travel again to Hong Kong, to be purchased, perhaps, by the expensively perfumed wife of a Texas oil magnate, something (for a time) to remind her of the trip.

The city of Djibouti, as distinct from the state of the same name, at the edge of which it is set, is a peninsula, a spit of land, a tongue lapping at and into the brine. Roughly speaking, the most affluent parts are those farthest out to sea. There is first the port area itself, the *raison d'être* of the whole, which folds forward protectively upon the ships. The port area forms the north-western end of the peninsula. Its north-eastern end, the Plateau du Serpent area, is given over to expensive residences (some enveloped in barbed wire from the seaward) – embassies and the like.

If the peninsula is thought of as a T, the port veers left and the Plateau du Serpent right. In the stem are located the most exclusive clubs – the Nautique, Aéro, Para, Hippique – as well as the ministries, better hotels, banks and shops, and the office of the President. At the top of the T's stem stands that institution which most effectively symbolizes the spirit of Djibouti, more than any of the various mosques and churches: the Casino.

The southern extension of the T's stem forms Djibouti's 'centre', its buildings visored in their arches, enveloped in shaded sidewalks. The sight is a picturesque one. But further south, away from the sea, a different sort of life emerges. Under the arches where the rue de Moscou and the rue de la Mosquée intersect, one detects, lying about, a wealth of goat pellets, inset upon a larger fecal bed, the fall-out of one's fellow men. As the formal city, the city of brick and mortar ends, one enters a world far more rudimentary. One may traverse it by car, but one may be absorbed by it only on foot. The men play cards on jerrycans, or they engage in a version of the game which in Ghana one calls *oware*. A woman lies topless on her bed with the door ajar, unable to sleep. Other folk, more fortunate, lie asleep outside in the shade on cardboard. Children aimlessly kick at balls, alternately through mud and dust.

People have a way of retaining their beauty, as well indeed as some *sense* of beauty, even in desperate circumstances.

Many people inevitably wander in and out of the city. But for the most part this population seems a pretty settled one. They survive, if tenuously. Without various forms of mutual assistance, it would be impossible. The pretty girl who, elsewhere, might serve as a secretary, is here reduced to prostitution. An ambitious boy, who might elsewhere be converted into a clerk, is recruited as a guerrilla or poacher. But for most, there is no recruitment at all. Only the help of relations, perhaps, or friends. The soil is such that no one may go and till his own garden. This population to the south of the city centre sets limits to the political prospects of the peninsula. Somehow held aloft since 1977, as if by magic, the government, despite its successes in the eighties, and facing a presidential election in 1987, can only hope that its audience, for long hurt by drought and unemployment and repeatedly regaled with tales of ministerial corruption, may not breathe too hard, or shout.

Djibouti has the air of a city which never really shuts down. While some sleep, others may work or chatter. Even in the early morning hours, curfews apart, little clusters of people are gathered on corners, sometimes quietly, sometimes noisily, pursuing hopes of something better, or at the least of something different. But everywhere around them at night are bodies, lying on their backs, doubled up in corners, laid high on army beds, laid low on cardboard mats, stretched out face down or up on bare cement. Everywhere there are bodies, of sleeping men and women and children, which, by the time the stores open in the morning, all disappear, leaving a pungent memento of their nocturnal presence.

Paradoxically, when Djibouti first achieved its independence, it was reckoned to be the fourth wealthiest state in Africa, after South Africa, Nigeria and Gabon. It is a country in which, despite high unemployment levels, to hire a maid or a gardener, by the standard of most African capitals, is nothing less than exorbitant. A refrigerated glass of mineral water, or a glass of sweetened and diluted lemon juice, or the most modest of meals, perhaps a cheese sandwich (*casse-croûte*), will strike any visitor from Zimbabwe or Zambia or Kenya as expensive. A hotel of quite poor standard in Djibouti can be expected to cost twice as much as in Nairobi.

Virtually every item consumed in Djibouti is produced somewhere else – Kenya (even the pineapples), Ethiopia, France, Somalia. There is such a press of people to feed and accommodate and amuse who do not themselves produce. At virtually every corner and passageway of almost any street of any length there is a bar or restaurant or hotel or brothel, all much of a muchness, with punchy names like Mickey-Bar, Ginn Fizz, Walla, Stop-Bar and the like. To provide amusement in Djibouti is big business, the sort of raunchy business associated with any major port, like Mombasa or Dar Es Salaam. But in Djibouti, these *divertissements* move closer to dominating all economic life than in any other maritime nation on earth.

In the highland states, the worker in a wet year can grow at least a modicum of food to supplement his income; even if a given individual cannot, others are usually able to supply his need at slight cost. Not, naturally, when famine settles upon the land, but famine does not conform to an annual cycle, any more than the drought which may trigger it. In Djibouti, there are no meaningful wet years to speak of. In sum, the worker has to be paid an 'exorbitant' wage, in foreign exchange terms, since otherwise he cannot live. If he or she has a job, it must also serve to sustain both relations and friends. It is a part of the code, and it is equally a form of insurance. Hence the peculiarity of high money wages, an impressive income per head of population, and pervasive unemployment, prostitution and the hint of a steely edge engendered by destitution. There is also, however, a certain quiet *camaraderie* among the poor, a certain cohesiveness and common understanding based upon an organic and socially undifferentiated suffering which one detects beneath the soaring violins and rhythmic thumps of American and British and French popular songs which appear to rend every cubic foot of Djibouti's warm night air.

The centre of gravity – politically, economically, demographically – of the Republic of Djibouti is its capital. It serves the highland trade, and the rest of the territory is so much space to move through. Djibouti is almost entirely surrounded by Ethiopia, except for a coastline of about 180 miles and the eastern border with Somalia, which extends for approximately forty-five miles. The territory is about 23,000 sq. km. or 8,350 sq. miles, about the size of Massachusetts or Wales. The altitude moves from sea level up to over 4,000 feet. The terrain is rough and almost entirely desert or semi-desert. Like Ethiopia itself, the land here alternates between flats and faults. Permanent vegetation is to be found in the highest areas, but these are of no economic significance to the country. The heat is enervating, even in the cool season, when temperatures average 86°F (30°C). When it is really hot, during June, July and August, the average temperatures exceed 104°F (42°C). The well-off lead an air-conditioned existence; every shop, bar, restaurant and hotel room in the middle money range is equipped with one or a battery of overhead rotary fans. The poor, who literally lie beyond the reach of such gadgetry, drily crumble; noontime, at its hottest, is shuttered and tomb-like, far quieter than the middle of the night.

France established her position in Djibouti with relative ease; she held on to it long and tenaciously; and she surrendered it only under duress. She regarded the position as too valuable to allow it to slip negligently into the hands of others. And it is fair to say that she still and understandably resists the logic of such surrender, even in the period which has followed independence. France had no more desire to withdraw from Djibouti than Britain from Gibraltar or America from Panama. The principle of self-determination is all very well, but we are still learning that strategic and economic self-interest is not a matter which any nation – imperial or otherwise – readily lays to one side.

The French first established contact with the King of Shoa, an earlier Selassie, on the Ethiopian plateau in 1839. To reach him, they travelled from the Djibouti coast. The advantage of Djibouti is the Gulf of Tadjourah, where the Gulf of Aden cuts deeply into an otherwise undented African coast, extending inland for 60 miles (100km.), across for 30 miles (50km.), and reaching to underwater depths of 900 feet (about 300m.). Within the Gulf, which Djibouti entirely envelops, there are three ports, two on its northern lip (first Obock, then Tadjourah further inland), and the last across the water to the south (the city – as opposed to the Republic – of Djibouti itself). The deepwater port is at Djibouti. Over time, with the increasing size of vessels and volume of trade, the port of Djibouti became identified with the territory as a whole. Djibouti provides an excellent harbour, one of the very best along the entire eastern African coast south of Suez, and obviously the most suitable for the transit of Ethiopian goods. That highland traffic alone – to the exclusion of Egyptian, Sudanese and Somali interests – really remains the overwhelming justification for the continued existence of the port.

In the usual imperial style, France concluded separate treaties with local notables, beginning in 1842. There was only a tiny population, for obvious ecological reasons, in these early days. But even then, this population was not entirely homogeneous. It was Muslim, and pastoral, and nomadic, but there was no large-scale organization encompassing its inhabitants, nor really – at the time – any reason why there should be. There were common cultural factors, but also different traditions of origin, clans, dialects, even languages. Basically, to the north and east of the Gulf of Tadjourah, the Afar-speaking people were established, while to the south there were the Issa, a Somali people of the Dir clan-family. The Afar and Issa/Somali languages are not mutually intelligible, although the pattern today is for the Afar to be able to understand Issa/Somali, while the latter have little grasp of the former.

The first French stations were established north of the Gulf of Tadjourah, and therefore among the Afar. France concluded a treaty of 'perpetual peace and friendship' with the northern rulers, obtained docking rights and facilities at Obock, and formally attached the latter in 1862. The same happened further inland (and along the Gulf) at Tadjourah in 1884. Issa Territory was finally taken over from 1885 and the building of Djibouti began, the French encouraging Issa nomads to settle there in order to service the new city and port. All these different 'protectorates' – it was the fashion to name them so – were fused in 1896. Thus the Côte Française des Somalis, or French Somaliland, was officially born. The territory was called Somaliland, not because it was exclusively Somali, but simply because the most numerous people in what had become the only really important port – Djibouti – were Issa-Somali. The logic of this situation persists. For 80 per cent of the population of the city of Djibouti today is still Issa-Somali, a figure which neither Ethiopia nor (of course) Somalia disputes.

Djibouti was first linked by the French to the Ethiopian heartland by a caravan route. This eventually gave way to the rail link, which, begun in 1896, reached Addis only in 1917, twenty-two years and 486 miles (or 841km.) after construction first began. Djibouti was designated, by the France–Ethiopia Treaty of 1897, as the '*débouché officiel*' of Ethiopian commerce. The Djibouti–Addis railway was the first to traverse Ethiopian soil. It remains the most important. It has been traditionally via this line that the Ethiopian produce of the highland interior has reached its overseas markets. Any threat to it is inevitably perceived as a challenge to Ethiopia's economic lifeline, and thus as a challenge to Ethiopia's continuing existence as a modern state. The Haile Selassie government, for decades, both required and decried this dependence, and latterly took steps to decrease it. Thus the logic of the Ethiopian relationship with Djibouti is awkward. On the one hand, Ethiopia's easiest and most natural option is to depend upon the port, and so to seek in some fashion or other to control it. On the other hand, Ethiopia is compelled to concede the port's independence and therefore to diversify her outlets, in so far as Djibouti, for Ethiopian national purposes, is far from being secure.

In the modern history of the region, the rationale of the struggle between France, Britain and Italy was centred in the concern of each either to establish control for itself or deny control to its competitors of Ethiopian trade outlets – whether to the Red Sea, the Gulf of Aden or the Indian Ocean. The power which first came nearest to establishing overall control of the Horn was Italy, when, in 1936, through direct military conquest, she acquired control over Ethiopia, in addition to an extensive coastal area conquered earlier. But the Italian victory was shortlived. By contrast, France, after 1941, continued in charge of Djibouti, sharing control of the line to Addis with Ethiopia, by an agreement reached between the two states in 1909 for the creation of a joint Franco-Ethiopian company. For a time, the United Kingdom, following on the defeat of Italy in the Second World War, acquired control over the Eritrean coast, the entire Somali coast, Aden on the opposing shore and huge tracts of eastern Ethiopia. (It was not until 1954 that Britain altogether withdrew from Ethiopia.)

As long as Ethiopia and France could believe in one another, as it were, there would arise no particular problem relating to Djibouti. But in the immediate post-war period, pressure for native representation in administering Djibouti affairs inevitably arose. The Italians were re-admitted under UN mandate to what had been Italian Somaliland, which, it was understood, would lead to independence for that territory. (Italy, even so late, still sought to extend the area of the territory she held. This is the chief reason why the border between Somalia and Ethiopia was not settled at the time of the former's accession to independence in 1960.) Britain, which during and after the war occupied a vast area, comprising most of the territory in which the Somali nomads lived, voiced support (through Ernest Bevan) for a united Somali nation. This was resisted by other powers because it was seen as a British imperial gesture, which in part it was. The Somalis took up the cry of a national unity for the Horn, based upon cultural postulates shared by Somali peoples. The French, in the meantime, were prodded into setting up a representative council of twenty members (Conseil Représentatif) –

evenly divided between locals and Frenchmen – in 1945. This *conseil* gave way in 1956 to a new territorial assembly (Assemblée Territoriale) with thirty-two elected members and a local governing council with restricted powers shared among eight ministers.

The French consistently resisted any move towards independence for Djibouti, which retained the name 'French Somaliland' up to July 1967. It was then replaced by the more cumbersome formula, 'Territoire Français des Afars et des Issas' (TFAI). When the Territory was first called Somaliland, this was only in recognition of the fact that its economic centre of gravity had shifted south of the Gulf of Tadjourah; from Obock, above, to Djibouti, below; and in recognition of the fact that its population was no longer predominantly Afar, but Issa-Somali.

The immediate post-war fervour for independence, therefore, was located among the Issa rather than among the Afar. In controlling the city of Djibouti, the Issa would basically control the entire territory, including the Afar who resided within it. There was and remains a distinction, as we have seen, between these communities; there are further distinctions internal to each of them; but the only way in which the French could perpetuate their own control was by playing upon these distinctions, and by favouring the minority Afar community (there are no reliable statistics covering these groups for the territory as a whole) at the expense of the more numerous and potentially powerful Issa community.

President De Gaulle visited the Territory in 1956 and again in 1966, followed by President Georges Pompidou in 1973, where their common and continuing concern was to rally the Territory behind the principle of continued French rule and to affirm, in Georges Pompidou's words, '*l'appartenance de Djibouti à la France*' – that 'Djibouti belongs to France'. Somalia became independent in 1960, growing out of a union between Italian Somaliland and British Somaliland. Mogadishu wished also to incorporate Djibouti into this extended political family – one reason why the French, finally, in 1967, substituted for the expression 'Côte des Somalis' one which would emphasize the lack of continuity with Somalia.

In the early 1960s the French were in charge of Djibouti. The new Somali state had emerged further to the south. The French were on hand because of the port and the railway they had constructed leading into the highlands. That part of the line which lay within their territory, however, ran pretty much exclusively through Issa-Somali territory.

The new Somali state increasingly encouraged opposition among the Issa to a continued French presence. The French were compelled, in the new post-war situation, to make their control more democratic or representative. But in doing so they ran the risk of losing it to the Issa-Somali, who, it was expected, would be rapidly absorbed by the new Somali state to the south. It was less the simple independence of Djibouti which threatened French interests than its absorption by the southern neighbour, whose economic position was more precarious, and whose financial exactions – if absorption occurred – would probably prove more onerous. French policy for Djibouti therefore was to encourage the growth of a dominant Afar political power base. France pursued this policy in concert with the Ethiopian government of Haile Selassie (as she was more or less compelled to do, if she were to escape simultaneous conflict with both neighbours).

With French encouragement, Djibouti voted by two to one in September 1958 to continue as a French dependency. The vote demonstrated a unity of purpose more apparent than real, and within two months the French suspended the Vice-President of the Assembly and several ministers, arrested two other ministers and dissolved the Assembly. What France was trying to do eventually came to appear quite impossible: to create a truly representative regime in the colony, which would protect French interests and, as a means to this, freely elect to remain a colony. The prospect of remaining eternally subject to foreign rule was repugnant to African opinion throughout the continent: to the OAU, obviously to the policy of the Republic of Somalia, to the Issa-Somali community in Djibouti especially, and even to activist Afar youth. The combination of all these pressures provided a basis for continuing tension and instability over the years. The local, French-supported government became increasingly untenable.

On the occasion of De Gaulle's second visit to Djibouti in August 1966, there was serious and bloody conflict between those supporting and opposing the Ali Aref government, which is to say between those who did and those who did not favour Djiboutian independence. The army fired upon demonstrators, killing four and wounding seventy. De Gaulle refused to address a riotous assembly. Anti-French demonstrations were unleashed in Mogadishu, favouring Djiboutian independence and ultimate absorption by Somalia. In response to these developments, the Ethiopian Emperor (in September) spoke of his country's interest in Djibouti, and even went as far as to characterize

Djibouti as an 'integral part' of Ethiopia. But he had spoken in similar terms when British and Italian Somaliland were approaching independence. The French arranged a second referendum, nine years after the first, in which three out of five voters were declared to favour a continuation of Djibouti's status as a French colony. To produce this result, the French leaned more heavily than ever upon the local population, rigorously controlled entry into the territory by the Issa-Somali – who, like the Afar, are nomadic – and in a variety of unorthodox ways managed to obtain a favourable result. But this result was quickly followed by renewed rioting, a piling-up of dead and wounded, all capped by the expulsion of many Issas suspected of harbouring independent inclinations. By September 1967, even the UN General Assembly voted in favour of the proposition that France should concede independence to TFAI (as the territory was called until independence). In May 1968, an attempt was made to assassinate Ali Aref, and, although he escaped, the life of his chauffeur was lost. Nine months later, in late January of 1970, grenades were tossed into a bar/restaurant – Le Palmier en Zinc – just off Place Ménélik, wounding sixteen whites. The French response to this unpleasantness was not notably indulgent, and the main, basically Issa, opposition party (the Ligue Populaire Africaine pour l'Indépendance, or LPAI) had repeated cause for complaint against the repression (and more) generated by the determined fury of the French.

If Djibouti opted for independence, it was clear that the road would prove a rocky one, for continued French financial support was vital to the survival of the new state, and that raised serious doubts about the genuineness of such independence. The easiest option for moderate indigenous leadership was to remain under French control. That option, however, became increasingly unreal, because of political developments in the continent which overwhelmed Djibouti, and which neither the French nor the local leadership (or both together) could contain. Thus moderate local leaders were gradually but inexorably prepared to assume responsibility for putting together an independent government. The French, as we have seen, at first favoured the Afar community, under Ali Aref. As Aref's position was eroded, they finally switched their support to the Issa-Somali community, under Hassan Gouled Aptidon, now President of the new Republic.

The pressures at work upon the French to withdraw were both domestic and foreign. After Somalia became independent, she encouraged a like independence for TFAI. As it appeared that this might not be achieved peacefully, in 1963 Somalia sponsored the formation of a clandestine party, the Front de Libération de la Côte des Somalis (FLCS), based on Mogadishu. It was the FLCS, most probably, which was responsible for the assassination attempt on Ali Aref in 1968, and the grenade attack on Le Palmier en Zinc in 1970. The FLCS was certainly responsible for kidnapping the French Ambassador, Jean Guery, in Mogadishu, in 1975, releasing him a week later in Aden in exchange for the release of the man (Omar Osman) who had been condemned to die for the attempt on Ali Aref's life and the murder of his driver. In December 1975, yet another attempt was made on Ali Aref's life, for which the FLCS was again held responsible. And on 3 February 1976, the FLCS kidnapped a busload of thirty French school children and drove them to the Somali border post of Loyada, where the FLCS commando (six men) was gunned down by marksmen especially flown out from France for the purpose.

Ethiopia was satisfied for France to remain in control of TFAI, as long as the latter could actually maintain order. But there were clear signs that this was not to be. The French, the Ethiopians and the Afar were disposed to maintain that Afar numbers constituted more than half of TFAI's total population. Somalia, along with the Issa leadership in TFAI, reckoned Afar numbers as low as a quarter of the total population. If the Afar were to remain dominant, they could best do so with French support, especially military support, and would require the assistance of Ethiopia, which the latter was ready to give. But from February 1974, Ethiopia entered upon a revolutionary stage in her development, which made matters awkward for the Afar leadership. It also helped to dismantle French confidence in the prospect of a weak Afar government supported by a powerful hinterland state newly won over to socialism (and the attractions of nationalizing foreign assets and institutions).

The situation in TFAI was being rendered increasingly unstable by the operations of the FLCS from Somalia, on the one hand, while the dependability of Ethiopia as a supporter of French interests (both in Djibouti and in Ethiopia itself) was being brought into question by fundamental domestic changes on the Ethiopian plateau. If France left in power a weak Afar government, one necessarily dependent upon Ethiopia, at a time when the latter was turning radically socialist, the danger was that TFAI might simply be absorbed by Ethiopia, as Eritrea had been (by stages) in 1952 and 1962. And if that happened, Ethiopia would be in a position to take over the railway, just as Egypt had appropriated Suez.

By 1976, France was openly withdrawing support from Ali Aref. In late March, two members of the latter's party defected to the opposition, a third in early June. His government suddenly found itself in a minority. During 9–10 July, there were violent confrontations between Aref and Gouled supporters, leaving ten dead and fifty wounded. Within a week, Ali Aref resigned as head of government, which was accepted by France the following day, while the French Minister of Overseas Territories, Olivier Stirn, characterized Aref's rule as 'a cause of the trouble in Djibouti' (*'une gêne pour Djibouti'*).

This prepared the way for the setting-up of a predominantly Issa-Somali government, which would enjoy strength within Djibouti and share a fundamental opposition to any merger with Ethiopia. There was of course the much greater risk of a Djiboutian merger with Somalia, but the danger to French interests from this were probably considered less.

To begin, although Gouled's dominant and basically Issa LPAI supported Somali theses, and in turn was supported by the government of Somalia, and contained elements seeking union with Somalia (like the FLCS), no independent government in TFAI could expect to gain much from such a merger. There is no basic economic tie between Somalia and Djibouti but only between Djibouti and Ethiopia and between Djibouti and France. Somalia is at least as poor as Djibouti, indeed is probably poorer, and so would have provided no economic support and would probably have constituted a drain on the economy. Somalia has a military government, which was not the case at the time of the merger between Italian and British Somaliland in 1960. A civilian government, such as Djibouti's, may be expected to display a marked disinclination to be taken over by potential military replacements. A Hassan Gouled LPAI government, basically constituted of Issa-Somalis, would, at independence, take the sting out of – by simply meeting the demand for – Somalia's demands for self-determination (meaning Issa-Somali control) in TFAI. Such a government, in turn, with a budgetary deficit of about two billion Djibouti francs (or roughly US $12 millions) per annum, would require continued French support for the foreseeable future. It would thankfully permit itself to be protected by the French military against all comers. In such circumstances, therefore, 'before independence' would almost equal 'after independence'.

None the less, the intervention of Somalia in support of Djiboutian independence severely diminished the likelihood of any directly colonial role for France in the territory. If the French withdrew, however, Ethiopia was bound to seek to protect her own interests, first and foremost against the prospect of Djibouti's absorption by Somalia. Just as Somalia began by encouraging and training the FLCS, so Ethiopia responded, in the mid-sixties, by sponsoring the Mouvement de Libération de Djibouti – the MLD. The MLD was based on Ethiopia, with an Afar membership as its core. Just as the FLCS was expected to support Somalian theses, so was the MLD (which never however achieved any genuine importance) expected to promote Ethiopian interests. In the end, what really mattered to Ethiopia was not that she would take over Djibouti, but at least that no unfriendly power should do so – such as Somalia. On 31 July 1975, under the new military government, Ethiopia, before the OAU, renounced all territorial claims on TFAI. Somalia, under French and other prompting, had little choice but to follow suit, which she did in a joint Franco-Somali communiqué in early January 1976. At independence, the MLD disappeared from the scene, while the FLCS was theoretically absorbed by the Djiboutian military. Glosses on this development vary. On the one hand, the Djibouti government is disposed to maintain that it actually absorbed the FLCS, despite the latter's wild and high-spirited past. Detractors, however, argued that the Gouled government became a prisoner of the FLCS, on the assumption that the latter assumed effective control of the army, or nearly.

After Ali Aref's resignation on 17 July 1976, an interim government was formed, headed by an Afar married to a Somali, M. A. Kamil. His problem was to turn over power to the Issa-Somali majority of Gouled with as little protest as possible from within the Afar community. Kamil had previously served in the Aref government, but had begun to move away from it in 1975, in response to divisions within the Afar community itself. Kamil was characterized as an Afar patriot, but he recognized that Djibouti, with its overwhelming Issa majority, could not be governed without Issa support and the tacit agreement of Mogadishu. Kamil was all the more aware that Aref's power base within the Afar community was being steadily eroded.

The reason for the erosion of Aref's support was that the French came to regard his government as untenable. The untenability of the Aref government was clearly demonstrated in the period after February 1976. FLCS commandos, having hijacked a busload of French school children, were wiped out in a devastating display of precision sniper-fire by French forces. The basic demand of the FLCS commandos had been for the unconditional independence of Djibouti. And as it was in aid of that cause that they had been cut down, there was a revulsion of feeling – shared among Afar and Issa alike – against the Aref government.

But there were other important considerations apart from mere French disfavour. Powerful forces within the Issa-Somali community favoured a 'greater Somalia'. If the French were to withdraw, given the inferior numerical position of the Afar (probably over 40 per cent of the whole), the Afar would presumably eventually be overwhelmed anyway. Moderate-to-conservative Afar opinion, therefore, in order to protect Afar interests, showed increasing favour for the matching idea of a 'Greater Afaria' (*une Grande Afarie*), which would represent perhaps a force equal to that represented by the hope or threat of Somali absorption. At the least, the Afar could split TFAI and control the northern ports of Obock and Tadjourah. Alternatively, as winners, they could take Djibouti as a whole. The Afar are far more numerous in neighbouring Ethiopia than in TFAI itself. Any successful conclusion to the business of setting up an Afar State would clearly imply the dismemberment of Ethiopia (from which Somalia, too, hoped to detach territory).

Normally, Aref would have been able to anticipate significant Ethiopian support. But the Ethiopian revolution had begun in 1974. Land reforms followed. And soon the most important of the traditional Afar rulers in Ethiopia, the Sultan Ali Mireh, who owned extensive lands along the Awash, was in open revolt against Addis Ababa, and solicited support within Djibouti, from Somalis, from Saudi Arabia and from Kuwait. Aref was not able to support the Sultan, nor were the French. Aref split the Afar community on this issue. But, at the same time, he could not depend on any meaningful Issa support either. And given the campaign of terror mounted from Somalia against his government, some rapprochement with Somalia had to be achieved, sponsored by the French, under the interim leadership of an honest but ambiguously positioned Afar – thus the appearance and disappearance of M. A. Kamil.

Moderate Afar leaders were not very different in their outlook from middle-of-the-road Issa leaders. The options, from either side, were very limited. Either Issa-Somalis promoted union with Somalia, encouraging in this Afar support for a Greater Afaria (in different circumstances, perhaps union with Ethiopia), or both would choose to maintain Djibouti's independence, which would somehow mean Afars and Issas working with one another in ways which the French, for imperial purposes, had never previously sought nor, perhaps, desired.

The internal Djiboutian conflict between Afar and Issa has nothing primordial about it; but it does reflect a basic divergence of interests. Those sections of the population occupying the north and west of the territory (basically Afar) would tend to promote the interests of these areas, if dominant. Those sections of the population occupying the east and south (basically Issa) would be disposed to do the same for their areas, if dominant. The great advantage of the Issa-Somali position is its clean demographic predominance in the only part of the Republic, around the port and along the railway, which has any economic importance. Thus, given any clear split within Djibouti between Afar and Issa, the latter community would inevitably provide a more credible base for an effective government. But although, in such circumstances, it would be easier for the Issa-Somali than for the Afar to rule, it would remain a most difficult task for either to rule alone. Were the Afar to try it, they would have little control over the port and railway line. Were the Issa to try it, they would run the risk of generating opposition towards them among at least 40 per cent of the population.

The logic of the position would basically appear to be that any government in power must rest upon a distinctive power base in one of the two communities, but also that it must try to associate with itself less nationalistic ('parochial' or 'particularistic' or even 'tribalistic') elements from the other community. It must commit itself to maintaining Djibouti as an independent state. The other community, however, that finds itself out of power, and which – in consequence of this – is economically poorer and politically weaker and thus full of grievances of various kinds, will entertain livelier doubts about continued independence, at least in the form in which it presently affects them. When the Issa community was out of power – under the Ali Aref government – the political spearhead of its discontent was clandestinely expressed by the FLCS, through its raids, kidnappings and support for union with Somalia. Now that the Afar community is no longer dominant, the spearhead of discontent consists in equally clandestine cooperation with foreign interests, whether with Ethiopia, on the one hand, or with the Arab states, on the other.

If the Afar community were not persuaded that its interests would or could be protected by a predominantly Issa-Somali government, the options were these. First, the Afar could simply be repressed, which would remove all options. But that is easier said than done. Second, the Afar could actively promote a splitting of Djibouti, with either Ethiopian or Arab backing, providing Ethiopia with alternative sea outlets through the northern ports of Obock and Tadjourah. (Paris, by March 1977, had assumed this to be the Ethiopian plan.) Third, the Afar could promote the detachment of parts of eastern Ethiopia, to be joined either with the whole or part of Djibouti, thus forming an Afar-dominated state. (This was the plan attributed to the Afar Sultan Ali Mireh, with his large following in Ethiopia, along the Awash River, but also in the Djibouti Republic itself.) Fourth, the Afar could promote a variation on the above scheme, not leading to independence, but to some species of confederal co-existence with Ethiopia. There were further possibilities, but the heart of the matter was relatively simple. If the Afar had cause to think that they were significantly excluded from the benefits which attach to power, they could move as a community, probably under some form of socialist leadership, to undermine the Issa-Somali government, and most probably in association with the Ethiopian government.

If the Djibouti Republic were absorbed by Somalia, the Afar position would be a most uncomfortable one. The Afar aim, therefore, was first and foremost to avoid such an outcome. This was equally the aim of Hassan Gouled Aptidon, the Issa-Somali President of the Republic. But Gouled was fully familiar with the FLCS's desire for merger with Somalia, and yet he integrated it into the Djiboutian military. He had no illusions about the Afar opposition equally falling back upon external powers, principally Ethiopia, to effect their own ends. The Afar parties (MPL, MLD and part of the UNI), for example, boycotted the Paris Round Table Conference of March 1977, which led to independence in June. It is true that the Afar leadership – Ali Aref and the MLD – counselled its followers to participate in the Djibouti referendum of 8 May 1977, and to vote in favour of independence, which meant to vote in favour of an Issa-led government, producing an overwhelming 99 per cent vote in favour of independence. But signs of disaffection remained. In the city of Djibouti (Issa country), 96 per cent of all eligible voters participated. In Tadjourah (Afar country) only 42 per cent of

eligible voters did so. Thus, although the Afar supported independence under Issa leadership, they did so in a manner singularly un-
enthusiastic.

Parties out of power conventionally excoriate parties in power for
instancing behaviour which the first lot would themselves instance –
were they actually in power. The Afar (Ali Aref) government was
repeatedly attacked by the FLCS and, in different ways, by President
Hassan Gouled's PA I, for supporting the continued presence of French
troops. These troops were of course on hand to defend against any form
of external attack but were equally conveniently at hand to promote the
purposes of the local government in power. Similarly, before and after
the Issa (Gouled) party came to power (May 1977), it was in turn
excoriated by Afars for supporting the indeterminate continuation of
a French military presence. The Issa leadership, when under Afar
government, complained that French troops were being used to keep
nomadic Issas out. And they were. The Afars, under the new Issa
government, complained that these troops were holding back Afar
immigration, and permitting unbalanced entry to Issas to reinforce the
position of the latter. Many diplomats in Djibouti in 1978, representing
conflicting national interests, argued that the Afar complaints were not
without foundation.

This was a real problem for the Gouled government. It was difficult
not to credit the continuing avowals made by President Gouled himself
and other members of his government to the effect that their chief aim
was to preserve the independence of Djibouti. He and his Directeur de
Cabinet, Ismail Guedi Hared, repeatedly conjure up an image of a
Djibouti modelled upon Hong Kong, Switzerland or (in better days)
Beirut. But the strength of the government is uncertain. There was
governmental fear, at and after independence, of the Afar opposition,
and concern about the lengths to which the latter might be prepared to
go to achieve their ends, or alternatively, to vent their frustrations.
When, on 15 December 1977, just half a year after the independence
celebrations, grenades were again lobbed into Le Palmier en Zinc
(killing six and wounding over thirty), President Gouled immediately
held the Afar Mouvement Populaire pour la Libération (MPL) responsible – the MPL representing the radical wing of Afar opposition.

Many Afar, as also the Ethiopians, state that the perpetrators of this
attack, the victims of which were basically French, were the FLCS.

They further argue that its purpose was to turn the French against the Afar, as also President Gouled, and so augment his dependence upon the Issa and upon Somalia. President Gouled certainly implied the involvement of Ethiopia in this incident. He ordered large-scale searches, which continued for a month, mostly within Afar areas. These searches and attendant arrests led to the resignation of five Afar ministers, including the Prime Minister, Ahmed Dini, who claimed that the government was oppressing the Afar. President Gouled however maintained that his reaction was the correct one, and that the true culprits had been apprehended and would be brought to justice. Diplomatic sources in Djibouti, however, suggested that this would never happen, simply because some of the people held pointed to higher-level involvement, which, if revealed, could trigger a serious political explosion in Djibouti.

In short, the Issa government in Djibouti – the problem would have been even more acute were an Afar element in power – occupied very shaky ground. The Issa-Somali government sought security, and basically against the Afar, but to push too far this tendency would only further alienate the Afar, and thus create precisely the result it sought to avoid.

A predominantly Issa government has remained in power in Djibouti. But this government, like the country itself, has and will remain largely dependent upon the port and railway. It has worked to maintain smooth relations with Addis Ababa. But it has also remained on good terms with Mogadishu. For the Djibouti government not to show some modicum of favour to the Mogadishu regime might invite the latter to attempt to destabilize it. The Issa-Somali of Djibouti, if Mogadishu so decided, might be encouraged to perceive the Djibouti leaders as dupes and worse, to create public disorder, with a view to a Somali merger. On the other hand, for the Djibouti government to be perceived as unduly intimate with Mogadishu could reinforce Addis Ababa in its determination somehow to reduce its use of the port to an absolute minimum or itself attempt to destabilize the government.

In the long run, the greatest need of the Djibouti government, whoever happens to run it, is for peaceful association with Ethiopia. And the wisest course of action has been for Djibouti to attempt to create that degree of distance between itself and Mogadishu required to achieve this outcome. The basically Issa-Somali government in power in Djibouti from May 1977 came to power largely because of the backing it received, in various ways, from Somalia. But Issa predominance is more or less natural and normal in Djiboutian circumstances anyway, because it constitutes an overwhelming majority located at the economic heart of the tiny state. The Ethiopian fear has been that Somalia might simply absorb altogether an Issa-led government or that such a government might too easily be manipulated by Mogadishu in a manner vitally inconsistent with Ethiopian interests. The objective must be for Ethiopia to have confidence in an Issa-led government so that the two may work comfortably in tandem.

The fact is that neither Addis Ababa nor Djibouti wishes to depend upon the other in the degree that a fruitful working relationship would dictate. Up to June 1977, 60 per cent of Ethiopia's imports and 40 per cent of her exports found their way via Djibouti. The railway was knocked out of commission in the course of the 1977–8 Ogaden war and was only restored after June 1978. There was simply no way in which Ethiopia or Djibouti could defend the line from Somali guerrilla attack. In the war that Mussolini launched against Ethiopia in 1935, Italy had access to the sea; Ethiopia had not. France had banned Ethiopian use of the Djibouti line for military supply. Britain had acted similarly in regard to the coastal areas which she controlled in the region. Ethiopia, understandably, after the Second World War, was concerned to ensure that history should not repeat itself. The Ethiopian treaty with France in 1959 was projected to extend to 2016. By this treaty Ethiopia was accorded full access to Djibouti, in and out of war; was exempted from all import duties, and from all customs inspection, whether of merchandise or passengers travelling via port and rail. Shares in the CFE (Chemin de Fer Franco-Éthiopian) were equally divided between Ethiopia and France, the French half, in turn, being subdivided almost equally between government and private shareholders. The company was managed from the Ethiopian capital, all substantial technical work being handled to the north-east at Dire Dawa, while nominal control of such work was centred at Djibouti. These arrangements have been under consistent review in the intervening period, except that they now feature the new state of Djibouti as interlocutor. Each side, as always, wishes to secure for itself the best possible terms.

The difficulty, as the Ethiopian side discovered in the course of the Ogaden war, is that paper guarantees against a guerrilla adversary, frequently repudiated by its state backers, are to no avail. Ethiopia lost all use of the Djibouti line for the greater part of the conflict and even when the line was re-opened after mid-year 1978, it took no more than a month for the WSLF to mount yet another attack, causing serious damage. The line, in any event, is not in mint condition, so that it does not take a great deal to render it inoperative. A derailment of an entirely accidental kind occurred in January 1985 causing 450 deaths. When systematic sabotage is thrown in for good measure, a problematic situation is compounded. On 17 July 1985, for example, the line was sabotaged in Ethiopian territory, 150km. north of Dire Dawa, nearing the Djibouti border, blocking all traffic on the line for six days. There is no doubt that, in periods of calm, Djibouti must be viewed as the economically rational port of preference for the Ethiopian hinterland. The problem is that there has been precious little calm in the region over the period following 1974, and even earlier. Ethiopia is compelled, accordingly, to try as far as possible to secure or maintain other outlets.

Ethiopia has consistently attempted to reduce its dependence on Djibouti. It has attempted to increase and diversify its routes to the sea. It has attempted to tighten control over the outlets to which these routes lead. New and improved highways to Dire Dawa and to Assab have been constructed. (The port at Massawa was mostly used for Eritrea, while the ports at Assab and Djibouti were basically used for the Ethiopian heartland.) The road from Dire Dawa has been extended, branching from the railway at the city, moving towards the sea at Assab. A road has also been constructed running from Dire Dawa, parallel to the railway, into Djibouti itself. Apart from this diversification of traffic, Ethiopia has been concerned to consolidate its control over all of its coastal outlets, whether by absorbing Eritrea, over the period 1952–62 (with considerable US support, both military and diplomatic) or by renegotiating, in 1959, highly favourable terms (with the French) for the use of port and rail at Djibouti. Her powerful partners may change but basic Ethiopian policy persists.

The Ethiopians are particularly keen to keep and to develop the Eritrean port at Assab. Assab, at least, falls under their sovereign control and they are determined to accord it priority over Djibouti. Ethiopia also redirected some of her shipping about 240km. across the Gulf to Aden, flying it from there into the highlands. Aden, however, is far too expensive an alternative for bulk transport. It is open to Ethiopia to encourage the Afar to rebel with a view to secession, so as to make use of Obock and Tadjourah instead of Djibouti. Quite apart even from an unlikely secessionist option, it would not be difficult for Ethiopia amply to repay a hostile Issa-dominated government in Djibouti. None of these options could be regarded as ideal or economical or, in the last cases, moral. That which Ethiopia has favoured, namely Assab, has of course been at the expense of Djibouti.

Ethiopia appears to think – to some extent it is correct – that it has the whip hand in Djibouti. This is to say that there are alternatives, however costly, to Djibouti. Were the extreme point reached where Djibouti were altogether dumped by Addis Ababa – the possibility is so distant as to be practically unreal – it might conceivably become a dead city. There is a model for this line of development. Zeila, about 50km. south-east along the coast into Somalia, in what was known as British Somaliland, used to be the point of origin of the major caravan route from the Gulf of Aden up and onto the Ethiopian plateau. Zeila was the principal *débouché* of this trade and thus an important town. What replaced it (a matter of international politics), what siphoned off its trade, shutting out prospects of further growth, was the preposterous French railway at Djibouti – which took so long and cost so much to build!

Djibouti ears tire of talk regarding dependence upon the interior, upon Addis Ababa. The terms agreed with the French, after all, do not appear to have yielded much direct advantage to Djibouti itself: no customs inspection, no import duties and so on. But these matters are subject to negotiation. And the terms on which Djibouti does its job must be improved. Revenue does accrue to Djibouti from carriage, naturally. And while attempting in every way to diversify, it cannot be in Djibouti's interests to see use of the rail system decline. The Ogaden war, in breaking the service, also cut deeply into employment and income. It is precisely because the line does matter to Djibouti that her leaders may not wish it to. They are moving, in concert with the Ethiopians, to modernize it, but there is hesitancy: the air is full of fears of war. In this regard, Djibouti's hesitancy is at one with Addis Ababa's. Just as the Chemin de Fer Djibouti-Éthiopien is not secure enough an asset to justify unstinting faith, neither is the port. And a great deal of money is going into the port. On 13 February 1985, a brand new container terminal was opened, with two giant cranes, which can off-load twenty-five containers per hour, placing it well ahead of Aden across the Strait, and ahead too of Berbera (Somalia), Mombasa (Kenya), Tanga and Dar Es Salaam (Tanzania). Despite this investment, the port cannot be an altogether safe bet in so troubled a region either. The 1967

Arab–Israeli war closed the Suez Canal until 1975; revenue to Djibouti from her port facilities entered into a steep decline; by the time the Suez was reopened, super-tankers had been introduced to ply the Cape route, being too gross to traverse the Canal. It will be plain that none of this did a great deal to boost Djibouti's recovery. So Djibouti will and must move to modernize her port, and the highland rail link, and she must and does seek to diversify in other ways still.

A donors' conference was held in Djibouti in November 1983, result-ing in the agreement to supply Djibouti with assistance to the order of $400m. Hence the new container terminal. Hence, too, a new dairy, opened in November 1984, which produces a considerable range of milk and other products, and which has pushed the mineral-water bottling factory at Tadjourah into second place in terms of productive value. Significant developments, following the 1983 conference, are envisaged for Djibouti, involving in 1985 costly renovations to the stock yard, abattoir, animal food factory and much else, with the commencement of drilling to locate geothermal energy reserves. Developments of this kind, however, depend upon very considerable development aid. But the reason why such aid has been provided stems almost entirely from the strategic location of Djibouti.

The war that erupted in 1977, coinciding with independence, augured ill for Djibouti's prospects. The railway was soon to become inoperative and the port seemed already to be slipping into senescence. But France's original concern for position at the Strait of Bab El Mandeb had not lost its hold, especially after the withdrawal of Britain from Aden across the water in 1967, to be replaced by the Soviets, notably from 1969. Issa concern, even that of the Afar, began to move away from the prevailing fixation in the early seventies of simply removing the French from Djibouti. France was herself effective in persuading her European partners of the common strategic sense of maintaining an effective presence in Djibouti. The new pragmatic leadership in Djibouti, aware of the precarious position of the new state and of its geographical significance, became determined that it would not be absorbed by Somalia, and was in accord with Ethiopia that Addis should not play the shark, and hence made a firm play for aid from the West. French forces were kept, with their armoured vehicles, tanks, fighter jets, all conspicuously engaged still in desert exercises, to discourage adventur-ism. The object and effect of the donors' conference of 1983 was to strengthen this presence, to make it credible to Afar and Issa alike, by joining it with moves to ensure substantial economic development. Djibouti, accordingly, appears overnight to have become, with important new telecommunications developments, the most important and secure Western base at or near the Strait of Bab El Mandeb.

Djibouti is not prosperous. Her nominally high $400 GNP per head is misleading. But Djibouti is different from her near neighbours, Somalia, Ethiopia and Sudan, in that, though affected by war and frightened by it, she has none the less managed to keep out of the fray. So what Djibouti does now have, so unusual in the region, is peace. This is a boon which people who do not know war are far too likely to undervalue. Mengistu was invited to visit, and accepted, so as to be reassured *in situ* of Djibouti's concern to maintain friendly relations. The two sides have been cooperative and business-like in their approach to the future of the railway. Moreover, the French have promoted this policy of cooperativeness, without any direct or indirect commitment to Somali irredentism, such as has been displayed by first the Soviets and then the Americans, and which has been a cause of a great deal of death both by famine and bullet in the region. French and European support for the Gouled government, as Gouled has remarked, is not because of his beautiful eyes! But enlightened self-interest, however deficient in moral fibre, is always to be preferred to Greene's perennial Ugly American (and in a different way his Soviet counterpart), who appears to find it impossibly difficult not to vaporize the other into the mythological realm of the Evil Empire.

The Ethiopian highlands, giving rise to the Blue Nile in its flow to the north-west, and to the Webi Shebelle to the south-east, represent a powerful force, impossible to obstruct – if not here and there to divert – in the unavoidable descent to the sea. The Ethiopian highland people need not flood the coastal plains through which trade requires they make their way. It is perfectly feasible and desirable that their flow be diked and dammed in the degree required so as not only to avoid the destruction, but also to promote the enrichment, of the coastal peoples with whom the lives of the highlanders are joined.

The simplistic notions of national independence and non-interference in the domestic affairs of states, which are matching notions, must be accorded formal respect. The fact of space travel, satellite emplacement, complex surveillance systems, the labyrinthine mesh of the international economic system, the cross-national web of professional groups, however, all suggest that national independence and non-interference represent no genuine, no substantive reality. It is interdependence that is here to stay. Nationalism, these days, represents less the solution than the problem. The most nationalistic countries in Africa, Somalia and South Africa have also been the greatest source of death and misery since the end of colonialism. Ethiopia has been harsh in its dealings with Eritrea, in part precisely because of the troubling sweep of nationalistic counterclaims against its integrity. There is no excuse for injustice. But not much justice will come out of intractable civil wars and irredentist violence, whether these arise from some medieval notion of the Other as evil or batten upon the myth of autonomy as practical policy. These conflicts, as shown time and again in the Horn, are all too often exacerbated by outside powers that are equipped with the means to do so, but devoid of the restraint not to. Every great power has its role to play and each is obviously engaged. The question concerns only how these states play: whether like the dull computerized thugs they sometimes are – shortsighted, vindictive and threatening – or like rational adults, tempered by a common humanity, sensitive to the effects their policies have upon folk whom they cannot claim as fellow citizens.

# Part Three  AID MEMOIR

## 13 Seeking Peace, Eating Ashes

Democracy is in large measure desirable because it enhances the potential for self-expression among citizens. A large part of the value of free self-expression stems from the fact that it facilitates political decision of a voluntary kind, excluding the resort to force. Discussion, however, cannot produce agreement on everything.

If free discussion happens to lead anywhere, it is because it can emerge from pre-existent accord of some kind. Democracy is not an arrangement that we can unqualifiedly will into being: in some circumstances it is just objectively more difficult to bring off. Citizens require a great deal in common if they are to operate a system of open decision-making grounded in popular consent.

The range of pre-existent accord necessary to the elaboration of more detailed political agreement is of a very limited kind in most African states. In this regard, European states, culturally, are highly homogeneous, by contrast with African states, which are highly heterogeneous. While it is true that some European states are multilingual (Belgium, Switzerland and Yugoslavia, for instance), for all practical purposes most are not. In the United Kingdom, France, Spain, West Germany and Italy, despite internal conflicts in the first three approximating to a minimum level of civil war, there is one national language understood everywhere by everyone. By contrast, in Ethiopia, Sudan, Chad, Uganda and Kenya, there are literally dozens of different languages spoken. (The Nigerian government officially recognizes 250.) Language is only one area of difference, but it is matched by very many others, as between agriculturalists and pastoralists, Christianity and Islam, not to speak of the yawning gap between the urban rich and veritable armies of poor folk (both rural and urban).

The existing frontiers, within which the pullulating diversity of each African state is crammed, were not created by Africans. They were devised and imposed by European powers, with a view to promoting European economic and strategic interests. These colonies were not ruled democratically, but autocratically, and when the colonizers withdrew, mostly in the 1960s, they left behind new states united only in their desire to see the back of colonialism. The colonizers left behind a challenge to create democratic rule, but no genuine democracy itself. They left behind a history, more than half-a-century old, of autocratic rule, but no sure foundation (certainly not the colonial army) on which even this could safely repose. Contemporary African states must, of course, assume their responsibilities. But the chronic, structural problems they face are by no means all of their making.

Most African states are geographically small. Sixteen of them, from Mauritius at one end to Malawi at the other, are smaller than New York state. (Hence the fact that one-third of the United Nations' membership comes from Africa.) But they are not all small. Sudan, the largest, is about a third the size of the USA. Over twenty (like Botswana and Kenya) are larger than France – on the whole, considerably larger. But most of even the smallest African states contain quite sharp ethnic and other internal divergencies. In Guinea–Bissao – on the West African coast, with an area a fraction larger than Sicily or the American state of Maryland – there are at least six major languages spoken. Thus, although it may be true that the 'average' African state is small, almost all of them contend with a vast internal diversity.

Africa's frontiers remain today almost exactly as inherited at independence. The exceptions are the merger of Zanzibar and Tanganyika in 1964, the mutual exchange of territory between Nigeria and Cameroon in 1961 (both became independent in 1960), the merger of the two former British and Italian colonies as Somalia in 1960. Otherwise, apart from Gaddafi's Libyan conquest of a huge chunk of northern Chad above the 16th parallel – an appropriation, moreover, recognized and legitimated by no African or other state – nothing has been altered. It is not because Africa is happy with her frontiers. So much of African authoritarianism and civil strife largely stem directly from these. Wars in Chad, Sudan, Ethiopia, Somalia, Zaire, Angola and elsewhere are largely explained by the pressure of distinct ethnic groups (what in Europe would be called 'nationalities') to control their own destinies. Each is *de jure* a national state, *de facto* a mini-United Nations. But Africa can see no way of significantly adjusting borders without pulling the temple down.

Just as American federation, following the revolution of 1776, was seen as a response to a continuing European threat, so is the territorial integrity of contemporary African states – not to speak of a projected wider union or unions between them – seen as a means of warding off re-colonization. Fear of Europe (and America) largely explains the impulsion to hold onto boundaries which were themselves imposed. The fear is not so much of a wider union, which is celebrated in principle, indeed is even celebrated ritually. The fear is that even existing entities cannot be held together. Africa's fifty-one states are very fragile. It is this that explains the fundamental agreement reached at Addis Ababa in 1963 and incorporated in article III and reinforced in article VI of the Charter of the Organization of African Unity – that all member states should observe scrupulously the principles of mutual non-interference, sovereign equality and respect for territorial integrity. The overriding concern has been, not with the domestic rights of citizens, but with defence against any renewal of external subjection. The OAU 1963 Charter envisages and encourages wider union, but insists upon excluding disintegrative developments that would lead to still smaller groupings. Hence the inviolability of Africa's borders has become something of a sacred cow. The independence movement, which began out of a concern to protect the rights to self-expression of an African citizenry, became bogged down in a usually excessive authoritarianism associated with the unreasoning defence of colonially imposed arrangements.

The vast internal diversity of African states is a major obstacle to the emergence of fuller and more vibrant democratic processes. Authoritarian regimes, one-party states, *coups d'état* and military dictatorships are commonplace. And yet these arrangements are always justified on the grounds that they are consistent with the wishes (which could not otherwise be expressed) and the interests of the people. Here two distinct concerns are in conflict. One is with political self-determination, which if permitted would allow for an independent Biafra, Katanga (now Shaba), Eritrea, Western Sahara (former Spanish Sahara) and the like. The other concern is with perpetuating state units as they are, within their present frontiers, so as to stave off the collapse of the African state system as a whole. Given the extraordinary range of internal African diversity, the great majority of African states are subject to significant patent and latent secessionist pressures. Armies, whether overtly in power or not, are necessarily close to it, because they are the means by which these fissiparous tendencies may be contained, whether by threat or (more rarely) by combat.

Present-day African armies, like present-day African states, were essentially created by the colonial powers, and indeed with the same purpose: to keep the territory or state together. The colonial powers were in place roughly from the 1880s to the 1960s – for a minimum of fifty years. Mobutu Sese Seko (formerly Joseph-Désiré) of Zaire soldiered for the Belgians; Idi Amin of Uganda, for the British (King's African Rifles); Jean-Bedel Bokassa of the Central African Republic (later 'Empire'), for the French (notably in Indo-China). And indeed these states have survived intact. Their armies have served as a glue of last resort, whatever atrocities may have been committed along the way. The greater the pressure for self-expression, including self-determination, the greater the role of the army. Armies do not only fight. At best, they serve as a model of order and discipline. At worst, they become an extortionate rabble. But over the diverse peoples of virtually every African state, one rarely finds (as one does in Botswana) an underlying 'accord' or basic cultural coherence that undermines the military, or undercuts their role.

The army then is a crucial and pervasive institution, even in African states not directly governed by one. Armies must be supplied with new, expensive, usually dangerous gadgetry. They must be more fully consulted, better fed and more highly paid. They consume an increasing percentage of the limited national budget, whether because they are in power and comprise their own constituency, or because of the need to buy them off, to redirect an omni-latent lurch for power. Julius Nyerere sent the Tanzanian army into Uganda in 1979 to overthrow Idi Amin, following the latter's attack and annexation of Tanzanian territory (30 October 1978). In the end, however, Tanzania's triumph, won at the cost of perhaps US$1 billion, achieved little that might not have been better won by the Ugandans themselves, if over a longer term. And in the light of the July 1985 *coup* that overthrew Uganda's Milton Obote, the decision to pit Tanzanian troops against Amin's was all the more questionable. But such considerations ignore the chief question, which is whether Nyerere, had he acted otherwise, risked being toppled by his own army. It is not a consideration, given earlier cases of mutiny and conspiracy, that even he could ignore. Amin, at the time of his initial invasion of Tanzania's north-west, was certainly riding a military tiger, a restive army. His mistaken intent was to calm his minions by carving up a neighbour; he ended up as a refugee in Gaddafi's kitchen; and even that did not last for long.

The conflict between demands (1) for the self-determination of peoples and (2) for the respect of the territorial integrity of existing states is hard to ignore and even harder to resolve. On the one hand, patent or latent resisters (Biafrans, Somalis, Eritreans, the UNITA fighters, *inter alia*) may become heroes perceived merely to be struggling to free their peoples. On the other hand, governments of the day may be regarded as compelled by a true nationalism to do the only thing they can to save the state from dissolution – crush the rebels.

Considerations of weight usually (not always) lie with both sides. Economies of scale are important and the dissolution of states is a genuine risk. But advantaged peoples who have managed to capture government cannot reasonably suppose that their self-interested authoritarianism, where it systematically results in either the neglect or abuse of other peoples, is a fit matter for celebration – or tolerance. In all this, it will be clear that there is much space on either side, government and rebel, for rigidity, dogmatism and self-righteousness.

The quality of the leaders will not greatly help, in view of the chronically contradictory demands they will be expected to meet. Africa is littered with failed leaders, even those generally recognized as 'good', largely because no one can be expected to do the impossible – to serve as the avatar of a genuine democracy and at the same time to preserve the territorial integrity of a monumentally heterogeneous state. The basic difficulties of Africa's largest and most populous states – Nigeria, Chad, Zaire, Sudan, Ethiopia – stem precisely from this impossibility. All of these states have been ravaged by civil war and subjected to military control. Where that control becomes relaxed and veers towards 'liberalism', it as quickly becomes corrupt. Where it becomes more severely disciplined, it does not usually cease to be corrupt.

Except for the initial expense of conquest, the European powers spent far less in arming and policing the continent than is now found necessary. The first reason for this stems from the vast gap in technology between the European and African sides at the time. The use of gunboats down the Nile (into Sudan), of Gatling guns to subdue the Ashanti (in Ghana), of aeroplanes to smash the 'Mad Mullah' (in Somalia) eased the task of penetration. All major European powers became involved in this process. The risk attending involvement by so many states was that of armed conflict between them in the course of the 'scramble'. The purpose of the Berlin Conference, which began in October of 1884, was to contain this risk.

The success of the Berlin Conference provides the second reason for the relatively limited expense incurred by colonial powers in 'pacifying' the continent: they agreed to set up 'spheres of influence' so as to avoid military conflict among colonizers. There were, naturally, skirmishes, most notably that between the French and the English in 1898 at Fashoda (now Kadok) in the Sudan for control of the Nile. On occasion these gentlemanly arrangements broke down altogether: the interesting form of bush war conducted by the Germans in Tanganyika in the First World War; the mainly conventional war carried by the British against Italian Ethiopia from Kenyan and Sudanese bases in the Second World War; the blistering tank battles in the North African desert between Allied and Axis armies in the Second World War also. But, on the whole, there was no need to disburse large sums in Africa in order to defend acquired territory against the designs of 'rival' powers.

These conditions have changed. South Africa is in general terms one of the most destabilizing forces in the continent. She has raided most of her neighbours, has armed dissident groups and has directed them against their governments (newly independent Zimbabwe, Mozambique and Angola are cases in point). The South African government, even more than most of the colonial governments before it (colonial Kenya, Algeria, Congo, Rhodesia and so on), genuflects to a racist deity. It finds itself altars apart from the rest of Africa – an Africa which opposed colonialism and promoted independence more out of a desire to terminate racial rule than anything else. South African racialism, then, is under threat. It has taken and recycled these threats; it is on a war footing, up to and including a nuclear capacity; almost one fifth of the budget (US$3 billion) is devoted to the military. Racist rule in South Africa has ceased to be credible; that is why so much has to be spent to defend it. In principle the army is developed to defend against external attack. In fact, its role is, and is increasingly seen to be, to pulverize domestic opposition.

The military budgets of other African states have shot up for not altogether unrelated reasons. In principle, the concern is to defend against external threats. In southern Africa, many states are under threat from Pretoria: Lesotho, Botswana, Mozambique, Angola and Zimbabwe all find themselves in this position. In northern Africa, Libya appropriated (and still holds) Chadian territory. Ethiopia was invaded by Somalia (1977). Armed dissidents in Ethiopia are supported by Sudan; Sudanese dissidents, by Ethiopia. Uganda invaded Tanzania (1978); and Tanzania invaded Uganda (1979). There is a great deal of fear and plotting and meddling on all sides. But notwithstanding the volume of such activities, it remains that the chief job of the army in African states is not to defend against external attack, but to keep the lid on domestically disaffected groups. This is not necessarily and actively what they are doing. But it is this essentially that they were created and are maintained to do.

If colonial rule ceased to be cheap, one had to take the game elsewhere, or play it in a more sophisticated way. Governments (in Europe) did, as often as not, resist colonial expansion – as into Sudan to worst the Mahdi – on the grounds that it would cost them dear. But then to engage the Algerians in the north, the Zulu in the south, the Ashanti in the west, the Mahdists in the east, was a very cost-effective way, at least up the First World War, of keeping metropolitan armies out of trouble at home, giving them something to do in less sticky terrain, much further away.

Although there was the expense of setting up governors and attendant administrations, there were also the vast profits to be made by private enterprise, as in Algeria, Ghana, Ivory Coast, Congo (Zaire) and South Africa. In fact, governments usually followed trade (as in West Africa). Sometimes (as in East Africa) they created trade to pay for themselves. There was much to be traded. There was the mineral wealth, the tropical produce, the cheap labour, leaving aside the question of strategic emplacement, which was the chief reason for creating port cities like Djibouti. But the very process of setting up colonial empire initiated developments that would undermine it. Africans learned from their colonizers – which is to say that they also learned how to struggle against colonialism. And this was so whether the colonies were designed for settlers, like Algeria and Kenya and South Africa, or were run by European bureaucrats, like Ivory Coast and Guinea and Nigeria.

Communications between different parts of the globe accelerated, including transfers of modern military technology. Then communist systems joined their capitalist counterparts in the global village. The collapse of *entente* among colonizers, the contraction of the relevant military technological gap between colonizer and colonized, involved an equalling-up of advantage, making the prospect of successful guerrilla war against colonial government conceivable.

The British struggle against the Mau Mau guerrilla insurgency in Kenya lasted for four years (1952–6) and cost £50m. sterling. It claimed the lives of over 13,000 people (less than 100 of whom were European soldiers and settlers).

The French fought against the UPC (Union des Populations du Cameroun) up to and beyond independence (1955–63), at great cost, not least in loss of life (possibly as many as 20,000 killed). The struggle in Algeria against the FLN (Front de Libération Nationale) lasted (in effect) from 1954 to 1962, required the maintenance of conscription, flung half a million French soldiers against a nebulous enemy resorting to guerrilla and to terrorist tactics, led to 25,000 French deaths and the demise of a quarter of a million Muslims, and ended up costing between US$5 and 10b.

The Portuguese learned nothing from these earlier conflicts and became embroiled in war in all their African colonies – in Angola (1961–74) in Guinea-Bissao (1963–74) and in Mozambique (1964-74). Portugal maintained an African army in excess of 140,000 men, required to enforce a regime of national conscription and allocated a steadily increasing proportion of the national budget to the military, the latter finally absorbing (from 1965) more than half of all budgetary outlays, most of it occasioned by the last decade of war in Africa.

In the more pervasively democratic climate of opinion following a great war that had been fought, at least ostensibly, to preserve freedom, the cost of maintaining a colonial presence, either actually (as in Algeria or Angola) or prospectively (as in Ghana or Nigeria), was an added reason to withdraw. But the cost of holding together and defending the new states would not diminish – on the contrary. Where the military had been few, their numbers would grow. And since there was no entente between West and East, their rivalries would enter significantly into the new equations of force – most importantly in the Horn and in southern Africa. Western support for South Africa, for example, has usually been justified in terms of that country serving as a bulwark against communist (meaning Soviet) expansion in the region. The competitive arming of Ethiopia and Somalia fits into the same pattern. In sum, we observe greater pressure on African governments, greater instability in governments, constant expansion of military influence, and often of foreign influence, up to and including the fighting of proxy wars, accompanied by the testing, in new war zones, of innovative military hardware.

It would be a great mistake to suppose that military intervention in Africa follows simply from the indiscipline or authoritarianism or greed of the soldiery. The earliest *coups* in the post-independence period, and most of the rest, have been spurred on as much by the avidity of men in business suits as by men in khaki – whether local merchants and bureaucrats or foreign agents and diplomats. Soldiers do not live in a vacuum. They reflect their milieu. One important reason why they do not stay out of politics is that they are constantly invited from all sides to commit themselves. Roughly, this happens for two quite different reasons. On the one hand, particular leaders, actual or potential, civil or military, act to advance the cause/fortunes of an individual or clique. On the other, *coup* leaders fear that others will make the same extra-legal moves if they do not take pre-emptive action themselves. Citizens, civilian or praetorian, are disposed to act in part because there is no perception of a single 'citizenship', but of variegated nations; the ready supposition is that one's 'own' people must either dominate or be dominated; there is rarely any resting place in between.

Ivory coast in West Africa and Kenya in East Africa have both enjoyed civilian government since independence (1960 and 1963 respectively); both have experienced far greater economic success that their neighbours; both have allowed far greater freedom of debate than most other states in the continent. But, equally, both have been badly shaken by plots and attempted *coups*. One of the most chilling and revealing cases is Kenya's in the period leading up to and following on from the death of President Kenyatta (22 August 1978). There is considerable evidence to show that a private army (the 'Ngoroko'), largely supported by public money, was covertly established, not to fight cattle rustlers as claimed, but to ensure by armed force that the succession should go to powerful figures close to the failing President. Only a fortuitous sequence of events appears to have saved the then Vice-President, Daniel Arap Moi, from a fate less than presidential. Had the latter delayed departure (by a maximum of thirty minutes) from his Nakuru residence for the safety of State House in Nairobi early in the morning of 22 August, circumstances would probably have fatally militated against his assumption of the Presidency later that day. As it happens, even after Moi was sworn in, the then head of the Kenyan army, Major-General J. K. Mulinge, was approached by a well-known politician to execute a *coup* against the new head of state. Mulinge declined and rang the appropriate alarm bell. The then Attorney-General, Charles Njonjo, claimed that, had Mulinge acted otherwise, Kenya would have come under military rule. Sadly, that was not the end of the matter, for in August 1982 there was a fully fledged *coup* attempt, with a very heavy death toll (officially under 200), almost certainly of a pre-emptive kind, with evidence again of significant non-military encouragement.

Governments are always a target of discontent, but nowhere more so than in Africa. If the cost of grain or sugar or education goes up, some will suffer more than others. In the countryside, the suffering may not be noticed; in the cities, there may be riots. In a poor country, there is always much legitimate cause for complaint. There is a great deal of illiteracy. Folk do not readily understand the 'dismal science' and affairs of state. Leaders who presume to grasp what the 'people' want supply their own judgement instead. Unemployment is rife. People will sell themselves, not always to the highest bidder. Destitution, actual or prospective, can generate intense loyalties. Folk line up. They line up by region or language or economic condition. Journalists call this 'tribalism'. Some political scientists call it 're-tribalization' – meaning that leaders generate 'tribes', up to a point, where it suits them to do so. One observer speaks of 'the economy of affection', drawing attention to an extensive nepotism – except that there need not be any genuine affection, or unduly close or even traceable kinship.

A population that is not ignorant and destitute does not so readily sell itself. A functional elite is struggling amongst itself to determine who shall ascend to the summit of advantage and power. If the system is 'stable' – if there is no substantial rabble whose support can be bought or coerced, preferment will come via parliamentary or other non-violent routes. If the system is 'unstable' – if members of the elite can step outside established institutions, to recruit the uneducated and unemployed, the desperado, the poorly paid soldier with big bullets in his fist – succession is achieved in a less genteel fashion. In Africa, the people remain the collective plaything of the ruler(s). The regalia of a 'big man' is the 'hanger-on'. The system is as weak as this last, who will 'be hanged', in the two senses of the expression, if he doesn't 'hang on'. Until significant inroads have been made into this pool of destitutes (the 'real country', which remains outside, looking in) competing elites always have them to reach out to. And indeed they usually have little choice. The role of the army in Africa is a function of the vast cultural diversity of these states, taken together with the desperate position of the bulk of the citizenry. Without significant development, which requires substantial long-term development aid from outside, the population explosion will continue, as will impoverishment, hence authoritarianism and militarization.

African states are more empires than anything else. And empires, if initially republican, as was ancient Rome's, readily cross over from making war to being governed by warriors. African states are either under, or are seriously threatened by, army rule. European empire in Africa could not simultaneously prove both democratic and viable. To the extent that the colonial powers or their remnants stayed on (as with the Portuguese in Angola and Mozambique, and whites ruling by *Diktat* in southern African enclaves, like Rhodesia) they were increasingly compelled to fight guerrilla wars. To the extent that the European powers found a cut-and-run policy feasible, if not always attractive, it is because withdrawal did not necessarily entail economic loss. In fact, the basic deal struck in practically every case, even where colonial power was opposed by force, was to accommodate independence in exchange for the protection of foreign economic interests. Thus, in most cases, the transition to independence was directly supervised by the controlling colonial power. Colonies, because of political ferment, threatened to cost more than could comfortably be managed. The transition was normally achieved by interposing a temporarily democratic mechanism (an electoral contest between rival parties). Thus was a significant political liability disposed of. Colonial rule was reasonably peaceful, despite its abuses, and indeed its racism. These colonies were constructed pretty arbitrarily, from a human point of view, but they were backed by considerable power based on the metropolitan centre. After independence, in most cases, the credibility of an indigenous central order began to crumble.

In the mere eight years of Idi Amin's rule in Uganda (1971–9), as many as 300,000 people were slaughtered; in later years a similar number of people perished. There is really no parallel to such concentrated bloodletting in the annals of colonial control. The slaughter under Amin is usually regarded as betraying the personal whims of a murderous buffoon. That this is grotesquely simplistic is demonstrated by the fact that these events continued under Obote, who was never so regarded. This is not to say that Obote ever killed anyone: he did not. The *killing* none the less continued. What was taking place was a struggle among the northern peoples of Uganda, and between these and the southerners (principally Baganda), for control of Uganda. It is, indeed, only because Uganda was originally put together by British colonial authority in the way that she was that such a sanguinary outcome became possible. The same holds for past civil wars in states like Zaire ('Congo-Leopoldville') and Nigeria, and for present conflicts in Sudan, Zimbabwe, Mozambique and Angola. The fact that the independent African government rules well or badly is neither here nor there. The point is that the conglomerate of people yoked together is so internally diverse, so impoverished and so ill-educated that, whatever a government does, there will always be some important constituency that is left out and which bears some grievance towards it. There is virtually no independent state in Africa whose government some element has not designed to seize by force, not even in tiny places like the Comoros or the Seychelles or São Tomé and Príncipe.

It is remarkable how few of these difficulties were anticipated by Africans in the period of transition from colonialism to statehood. Published colonial opinion is pretty consistently pessimistic. Published African opinion is consistently optimistic. The reason is that Africans were fairly uniformly held to be racially inferior, the ethnic diversity ('tribalism') of the continent being taken as supportive of the denigratory thesis. Africans in turn formally dismissed the fears of intra-African conflict as colonial propaganda; they not only *sought* unity, but thought there really *was* unity (beneath the surface differentiation). Africans, at independence, were generally not prepared to see a problem, apart from colonialism itself. In the eyes of the West, Africans were as one. Africans themselves initially acceded to this myopia. And Africans who did not were commonly derided as colonialist lackeys.

Africa is only now just beginning really to comprehend its own variety. It is an old, highly complex world, and the failure to recognize this diversity, and its legitimacy, provides a major breeding ground for gross intolerance. It is unfortunate that African states should exist within the arbitrary frontiers that they have inherited, and yet there is very little in the immediate future that can or should be done about this. Most of the solutions would only compound existing difficulties. There can be no retreat to earlier solutions, to colonial rule or to smaller pre-colonial systems.

After the Second World War it was no longer possible for colonial regimes to stay on. The experience of France in Algeria and Cameroon, of Britain in Kenya, of Portugal in Guinea-Bissao, Angola and Mozambique, all indicate that the escalating cost of containment would have been feasible only if the metropolitan system itself had become more highly militarized and authoritarian. The struggle to maintain colonial authority in French Algeria, for example, pushed France to the right; sparked the 1958 *putsch* which overturned the Fourth French Republic and brought General De Gaulle to power; led to arbitrary arrests, torture and near civil war in the metropolis; provoked the settlers to revolt even against their hero, De Gaulle (for nine days in January 1960); raised the spectre of French paratroopers swooping down upon Paris to overthrow the government; and gave rise to an OAS (Organisation de l'Armée Secrète) supported by four French generals and resorting to terrorism in Paris as well as in Algeria.

The colonial regimes quit, not so much out of the goodness of their hearts but because they found it either necessary or prudent to do so. They could no doubt have remained, just as the United States could have persisted in Viet Nam through the 1970s, had it been possible to persuade or intimidate enough people at home to back the escalation of violence abroad. The consequence would have been, for both home and overseas territories, an upward spiral of racism, police brutality and military atrocities; a burgeoning of concentration camps; and a firm, phantom-like retreat from hopes of stability. The syndrome, in short, would have been South African.

Up to 1985 South Africa represented a supreme attempt to retain colonial rule on a permanent basis. From 1948, with the victory of the Nationalist Party at the polls, an attempt more concerted than ever before was made to check black advance, by means of systematic residential segregation (the Group Areas Act); by ensuring inferior education for Blacks (the Bantu Education Act of 1953 and the Separate Universities Act of 1959); by banning African political parties; by even denying citizenship to Africans on the perversely inverted argument that the natives, not the colonists, were the real 'aliens' (hence the so-called 'Homelands', occupying a mere 13 per cent of South African territory); by eroding civil rights for everyone, providing for detention of suspects for ninety days (renewable) without trial; by creating an unrestrained security organization, given to letter bombs, 'suicides' from upper floors and 'dirty tricks' of various kinds, an organization appropriately styled BOSS (Bureau of State Security), whose operatives, ironically, were paid for by surpluses generated by black labour. To maintain colonial control *in situ*, South Africa judicially executes at least one hundred prisoners per year, keeps another 120,000 behind bars and in the first nine months of 1985 had the dubious distinction of shooting down over 600 people, mostly children, in order to quell their restiveness. To control local populations, South Africa has also to control neighbours who might harbour them, and their grievances. Hence the refusal of South Africa to accommodate, even under pressure from the Western contact group (led by the USA), meaningful independence for neighbouring Namibia, which South Africa has held since displacing the Germans in 1914, and held illegally since the UN General Assembly decision of 1966 and the International Court of Justice opinion of June 1971. Hence, too, South Africa's continuing attempts to subvert nearby governments in Angola, Mozambique and the Seychelles. Hence South African armed attacks on Maseru, the capital of Lesotho (which South African territory completely surrounds), in December 1982 and on Gaberone, capital of neighbouring Botswana, in June 1984.

South Africa is colonial government gone mad. But 'go mad' is, soberingly, just what colonial rule, on the whole, declined to do. It is because authoritarian minority rule does not work very well in any 'open society' that the colonial powers withdrew. Colonialism served its purpose, both for good and for ill. Had it remained, it would only have become a grotesque parody of itself, as in the South African case. The greatest tribute that can be paid to it, in the African context, is that, on the whole, it withered away, causing a minimum of pain. What it left in its wake was a massive poser which could only be resolved, if at all, by independent states, and by a smaller number than might be dictated by the thousand or so indigenous African languages, or by the even larger number of pre-colonial African political entities. The problem was to create new, stable polities, of mind-boggling human complexity, with wiltingly slender resources, on a sound democratic basis. No other region of the world, at any time in history, ever confronted a puzzle of such intricacy as this, devoid of meaningful models or reliable paths to follow.

Opting out was never really possible. Reverting to a smaller scale would not help, despite some supportive rhetoric to this effect. The new African states had come into being in a tightly unified world. To attempt to ignore that would simply invite more trouble. Smaller entities had already been wiped out. And now that Africa's population was spurting ahead, difficulties of scale and disappearing land raised their heads with alarming insistence. Africa was not to be allowed the time of other continents to come to grips with the difficulty. Twentieth-century imperatives had caught up and would not go away. A weakened Europe, not far from the close of the Second World War, sought to distance itself from the Pandora's box that an earlier, more inventive, self-righteous and expansionist Europe of yesteryear had created. An independent

Africa, inheriting a bog of meaningless boundaries, rising expectations, simplistic understandings, complex cultures and remarkably procreative populations, had nowhere to run. The adjustments required are proving quite as painful as the most pessimistic of well-meaning observers could have imagined.

It is important to observe just how peculiar Africa's difficulties are. We may take it that the general pattern of population growth is for hunting and gathering peoples to maintain low population density and low fertility. Agrarian peoples (who have been about at most for perhaps the last 10,000 years) produce more, live in more compact settlements, and hence reproduce more fulsomely. The great divide emerges between agrarian and industrial societies. At the outset of industrialization, populations swell egregiously, then, with greater productivity, better distribution and greater security, fertility declines. In the case of Europe, as agricultural land was enclosed and production accelerated, a massive surplus population was created. It did so in the infancy of the modern world, and famine did not absent itself from these developments.

But far more important than famine in controlling European population growth was new land, land still occupied (largely, not exclusively) by hunters and gatherers (in the Americas and later in Australia), peoples who, with a minimum of difficulty, could be removed from it by more populous and powerful agriculturalists. The agrarian populations of the East, of China and India most notably, were already thick on the ground, not readily displaced. This was not the case in the Caribbean, in North America, in most of South America (especially in Argentina and Brazil) or at the Antipodes. European overpopulation, in short, during the transition from agrarian to industrial systems, was managed in significant degree by means of settlement and colonization of new lands elsewhere. Western Europe is a singularly fertile environment, and this counts for much. It is a region that has never known famine in the seasonally recurrent fashion common to India and China up to about a generation ago. But what counts for at least as much as good land, and probably more, is the fact that Western Europe (most especially the British Isles) was able to export its human surpluses elsewhere on a massive scale. The Eastern empires, whether from inability or disinclination, did not do this, and so their history is markedly different.

As for Africa, in the period following European contact and the hotting up of slaving to provision the plantations of the New World, population declined. Carr-Saunders, a former Director of the London

School of Economics, estimated that the populations of Africa and Europe, around the year 1600, were roughly equal. The effect of slaving was to diminish population size. The effect of colonial conquest, over 300 years later, was to stabilize it. The spark of independence had the effect of setting up a daunting rate of growth, at par with the European rate of previous centuries. The Egyptians found themselves in this predicament well before the rest of Africa, and were much disposed to look to the south, to Sudan, to the upper reaches of the Nile, as regions into which they might send their people. These were vain imaginings, but entertained all the same. The rest of Africa cannot even comfort itself with this. There is nowhere to go, and little thought that there might be.

Africa is a population pressure point; everyone is anxious for land that is no longer there. Plural marriage is common, fertility high, life span brief, subdivision of land unremitting. Just as the land frontier of the American West and of the Soviet East has been exhausted, so in Africa, but with consequences here infinitely more dire. Africa's capital is People, her religion is Family, but, in the contemporary world, room for either is constantly diminished. A capital like Lagos or Nairobi, a port like Douala or Mombasa, magnetizes people for hundreds of miles around. No land being left, they go to the cities, where many are increasingly entrapped as thugs and prostitutes. The exhaustion of land in the country directly leads to crime in town. As land becomes more valuable, because scarce in relation to population, so may we expect more conflicts (civil wars) within African states over the allocation of resources to different regions. Biafra, Chad, southern Sudan, the Ogaden were/are all in part struggles over the allocation of increasingly scarce resources between the regions of different African states. Increasing militarization is a reflection of these difficulties. It is an attempted solution. But by appropriating scarce resources that might otherwise be directed to development, it merely compounds the problem. War, without appropriate intervention, can only be expected, over the coming years, to spread and to intensify.

# 14 The African Community: Retreat from Disengagement

African and other Third World states have important economic, political and cultural ties with the outside world, and most especially with the West. These states have moved from colonial dependency to political independence, reflected in the growth of UN membership from fifty-five states in 1946 to 159 states by 1986. These evolving relations systematically favour the West as opposed to Africa and the Third World. African states are primary producers, export one or a very few products and are adversely affected by fluctuating prices, which, in real terms, are deteriorating, matched by seriously accelerating debt.

It has been contended that all men, and all states, are caught up in a struggle for power after power.[1] It has been thought by many that a striving for independence, understood as autonomy, is a characteristic of all humans. Locke was of the opinion that individuals, discontented with their lot, could always pick themselves up and move off to other climes and lands – the world being conceived by him as a vast and rather empty space.[2] The notion that humans are greedy, aggressive and expansionist has in part been referred to as 'possessive individualism'.[3] Much has been said about the puritanical combination of individualism, acquisitiveness and self-denial.[4] These qualities certainly did nothing to obstruct colonial expansion abroad.

Whether or not all humans, at all times, are doomed to seek for power after power, some such striving underlay the expansion of Europe in Renaissance times. The small European stain on the global checkerboard suddenly spread, in the short space of 350 years, from the sixteenth to the nineteenth centuries, and enveloped the whole. Before the outbreak of the First World War, the only Third World states which had not been overrun were Ethiopia, Afghanistan and Thailand.

1. Thomas Hobbes, *Leviathan*, London, 1651.
2. John Locke, *Two Treatises of Government*, London, 1690.
3. C. B. Macpherson, *The Political Theory of Possessive Individualism*, Oxford, 1962.
4. See, for example, M. Weber, *The Protestant Ethic and the Spirit of Capitalism*, 1904–5, trans. T. Parsons, 1930; E. Troeltsch, *The Social Teaching of the Christian Churches*, 1912, trans. O. Wyon, 1930; R. H. Tawney, *Religion and the Rise of Capitalism*, London, 1926.

Others – Japan and China, for instance – were later to reassert themselves, but the pattern was clear. The effects of such imperialism have marked us all, and while some of them may be reversible, such as unbridled exploitation, racialism and gross inequalities, there are others – such as global integration based on massive technological innovation – which most certainly are not.[5]

Consider the various forms of contemporary integration. There is the United Nations system itself, with all of its constituent agencies – the World Health Organization, the United Nations Educational Scientific and Cultural Organization, the Food and Agriculture Organization, the United Nations Development Programme and the International Labour Organization. The United Nations Environment Programme is head-quartered at Nairobi. The Economic Commission for Africa is head-quartered at Addis Ababa. Apart from these institutions there are others of vital global significance, such as the World Bank, the International Monetary Fund, and the General Agreement on Tariffs and Trade, all agreed or foreshadowed at Bretton Woods in 1944. There is, too, a great range of other groupings such as the Organization of African Unity, the Arab League, the Non-Aligned Movement, the Group of 77, quite apart from a host of less extensive regional organizations to do with Africa and other parts of the world, all of which groups none the less work with and some even within the United Nations system.

Apart from considerations of supra-national organization, it is of course possible by ordinary commercial means to fly to virtually any part of the world within a day, to telephone or cable virtually anywhere within minutes or hours and to be advised by satellite almost instantaneously of events as they transpire in diverse parts of the globe. The world has been economically interdependent for a considerable time. There has been a shrinkage of its space by all manner of technological devices unparalleled in previous times – the telegraph, telephone, train, road transport, jet aircraft, rocketry, satellites, computers, complex retrieval, printing and copying procedures, the silicon chip, cinema, television and so on. The pace of change in all such matters, thus, has vastly accelerated.[6]

There is nothing new in this description of accelerated technological change. The effects of the Industrial Revolution have been remarked

5. See M. McLuhan, with Quinton Fiore, *War and Peace in the Global Village*, New York, 1968; and J. Millar, *McLuhan*, Glasgow, 1971.

6. A. N. Whitehead, *Science in the Modern World*, New York, 1925.

upon time without number. Increasing integration of world markets both preceded and gave added impetus to such change. But all the international institutions elaborated in this century merely reflect the conscious political need, both felt and objective, to cope with the remarkable range of difficulties created by these new developments. International institutions are only in their infancy. Those that we have are imperfect, to say the least. But their elaboration reflects more compulsion than impulse. The menace of novel weaponry is not the least crucial of many factors which induce national rulers to move, however haltingly and with whatever reluctance, in the direction of a globally collective security scheme. This movement reflects a complex evolutionary pattern featuring an unprecedented speeding-up of technological innovation combined with the growing impingement of distinct societies upon one another. It is perfectly possible in the twentieth century, as never before, for human beings entirely to destroy the world – whose origins their religions and myths have almost universally attributed to the indulgent fancy of one or some combination of deities. The United Nations system, like the League of Nations which preceded it, has less to do with 'idealism' than is conventionally supposed, and much more to do with a simple matter of vital and global problem-solving. In all of these matters, the process of global integration – I do not say homogenization – is irreversible.

In today's world, then, virtually no state is entirely independent of any other. There is mutual impingement at frontiers. There is interdependence within the global economic system. There is association at the United Nations and in its specialized agencies. There is a global system of communication. There are of course gross disparities in technology, power and influence among the world's states. As much as 90 per cent of the world's industrial capacity is located in the rich states of the 'North'. Such inequalities make some peoples readily and obviously subject to manipulation by others. But this situation holds as much within states as between them. Just as solutions have been and are being forged domestically, so do we observe a parallel phenomenon in the international arena.

Obviously 'continental' economies such as those of America, the Soviet Union and China are more readily able to move in the direction of autarchy. The external percentage of their total trade is in each case small – but also growing. China is, in every important sense, a Third World state, only her size, vast population (like India's) and political and cultural homogeneity (unlike India's) rendering her unique among them. The USA and the USSR, however, enjoy positions of dominance. Although better able to move in the direction of autarchy, they are doing precisely the reverse.

The USA, in alliance primarily with Western European states, has been the globally dominant political power since the Second World War. That dominance is now being challenged by the USSR, given its emergence in the 1980s for the first time as indisputable contender for the title of the world's most powerful military state. The systems of the First and Second Worlds, while being in a position to manipulate Third World systems, are also best positioned (in terms of available resources) to help them. The West benefits most from her relations with the Third World, including Africa, and there is no absolute way in which this embrace can be escaped.

Engagement, whether economic, political or cultural, will certainly not go away. The only means of dealing with inequities in contemporary global arrangements is a collective African and Third World response which cannot be entirely consistent with the promotion of simple national independence and autarchy. In the degree that inequities are overcome, the means of doing so lie not in disintegration, but in improved African and Third World cooperation. The integration – in the areas of communications, global ecology (acid rain, nuclear waste, destruction of rain forests), technology (including the military), culture, economy – will persist and even accelerate, whatever else happens.

'Modernization', 'growth', 'development', naturally enough, will mean many different things to as many different observers. But one of the most important implications caught up in these notions is the idea of equalizing power, in some degree, between rich and poor states – of somehow rearranging domestic and international systems in such a way as to make it possible for the poor to exercise a greater degree of control over their own lives in future. The historical charge that the West 'underdeveloped' Africa is in a significant and obvious sense correct.[7] But this is only another way of saying that the West invaded, shaped and exploited Africa for its own purposes. 'Development' still relates to an aspiration which obtains in the real world and which has to do with altering it. Past systems, for ill or otherwise, have in fact been destroyed for good. The only way to put the pieces together again is by going

7. W. Rodney, *How Europe Underdeveloped Africa*, London, 1972.

forward. To overcome historically cumulative inequities is not impossible. The achievement of such an effect will inevitably hurt certain interests in the West and in the Third World. But there are various measures which can be pursued – liberalizing trade, providing greater aid, stabilizing commodity prices, stabilizing arrangements with multinationals, increasing investment in Third World agricultural and industrial development – which should in fact prove mutually beneficial to Africa and to the North.[8]

'Development', in the central sense suggested, projects an egalitarian goal. This need not mean that equality is the only legitimate value in a wider scheme of things. But it is the dominant value at stake in international political and economic relations between rich and poor states from the African perspective.

It is at least misleading to suggest that the very *concept* of development performs some magically disembowelling effect upon the Third World.[9] It is true, as suggested, that there is a 'rigid power-stratification system' on the international level. It is to be doubted however that the propagation of the idea of development has enabled 'the elite nations to save themselves from isolation'. It is doubtful that a mere idea – 'development' – could achieve this effect. It is the less subtle fact of sheer power preponderance which has of itself irresistibly enveloped the rest of the world. On the whole, those systems which proved technologically most vulnerable *vis-à-vis* Renaissance Europe were those which were made, and in varying degrees continue to be, politically dependent and economically deprived. The initial gap and its contemporary manifestations are inextricably intertwined.

To accept the notion of 'development' – in the sense given it above – does not necessarily orient the individual African states more towards the elite nations than to one another. In fact, if the developing states do not organize themselves, for example as producer cartels, after the manner of OPEC years ago, it cannot be expected that the rich states of the north will make concessions with sufficient speed or seriousness. Irrespective of ideology, Third World primary products must be paid for at an economic rate, in stable conditions, and long-term loans and investment must be directed to the Third World both by the First and Second Worlds. The aim must be to avert ecological and demographic

8. See *The Report of the Independent Commission on International Development Issues under the Chairmanship of Willy Brandt*, London, 1980.

9. See C. Ake, *Revolutionary Pressures in Africa*, London, 1978.

disaster for the world as a whole. It is not inevitable that the world should be overwhelmed by events like the 1975–9 implosion of Kampuchea under Pol Pot (with a death toll stretching, by conservative estimates, to four million people); the tragedy of the Vietnamese boat people; the dramatic but less serious outflow of Mexican migrant labour into the USA; and the unprecedented scale of famine in Ethiopia, spreading to Sudan and other African states.

Third World organization, with its attendant pressure – most importantly upon the West, but also increasingly upon the East – is vital to progress. Nuclear Armageddon and ecological collapse are distinct twin dangers. The one directly threatens the destruction of the earth as a habitable planet. The other indirectly but cumulatively threatens the same effect through rapid deforestation, desertification, pollution, impoverishment, population explosion and the pile-up of massive refugee populations, which in Africa alone total perhaps ten million.

It is inevitable that war in some degree must attend such misery, with increasing pressure on wealthier peoples by those who are poorer, but with the most immediate conflicts among the poor themselves. Troubled by gross domestic crises, weak governments may seek internal support by means of external adventurism. King Hassan of Morocco, for example, saw his popularity at home soar in the attempt to take over Western (ex-Spanish) Sahara. But this endeavour of Hassan's – jointly pursued with ex-president Moktar Ould Daddah of Mauritania – met with fierce local resistance (following Spain's withdrawal in February 1976), produced the overthrow of Daddah (10 July 1978), deeply alienated Algeria, and brought the French and the Americans in on the Moroccan side to prop up a Westward-leaning regime. President Siad Barre of Somalia, in far more desperate domestic circumstances than King Hassan, launched his army (1977) into the Ethiopian Ogaden, with ambiguous encouragement from the US government of the day. The USA was concerned to undermine a leftward-leaning Ethiopia. This spurred the Soviets on to an enormous effort to salvage the position of their client. In this they succeeded (1978), demonstrating their airborne ability to deploy rapidly and massively to distant parts of the world.

Third World proxy wars between the super-powers, the exacerbation of the arms race between the latter and between their clients, can only increase the probability – as threatened in the Middle East, Ethiopia, Afghanistan, Iran and elsewhere (for example South Africa) in the Third World – of miscalculation leading towards a direct conflict between the principals. This could be on Third World soil by means of limited tactical weapons or globally by means ultimately of strategic weapons – with their uncontrollably destructive power. Africa, along with the rest of the Third World – heartily aligned as so many of these states effectively are – has none the less a clear interest in avoiding such developments and in sustaining a genuine non-aligned movement as a means to survival. The Third World's interest in this, as it happens, would serve everybody's interest.

The danger with each super-power is that it enjoys the dubious luxury and great risk of hearing only its own counsel. If each only speaks to the other, each will perceive that what it hears is merely the brief of an adversary equipped with equal resources. This super-power condition of being judge in one's own case is a precondition more for conflict than for conciliation. The very fact that the Third World has severely limited resources makes its striving for political independence between these camps all the more imperative. Different and less committed views require to be aired. But this cannot be done by individual Third World states: they are far too weak to attempt to act in isolation. The most substantial leverage they can hope to achieve will follow from acting in combination, in the teeth of super-power determination (which will wax and wane) to compel them to choose sides.

Most developing African states, whatever the gulf between theory and practice, none the less seek some reasonable balance between the imperatives of regional cooperation, on the one hand, and improved economic and political relations with the wealthier northern states, on the other. Without intra-African cooperation, the ability to exert greater pressure for increased support will – in general – be much diminished. The problem with the present global system is its inequitable character – especially *vis-à-vis* Africa. To counteract that requires greater attention to domestic inequities, to regional, continental and to Third World cooperation generally, not as opposed to improved north–south arrangements, but as a means of achieving them.

African states may withdraw from heavy investment in expensive and dangerous imported drugs, elaborate hospital complexes, specialized surgical teams (as for open heart, transplant and similar operations), a national airline for each state, the importation of luxury items (such as expensive cars), impressive but wasteful urban construction schemes, fancy but inessential machinery and so on. It would be better and cheaper most often to encourage preventive medicine, breast-feeding, rural clinics, intermediate technology, labour-intensive techniques, to pool airline and similar services on a regional basis, and to raise tariffs sharply on luxury goods of every description. But these would only be means of drawing back to leap further. To implement such measures requires not only courage and probity, but also an investment in education and the improvement and creation of skills, and one of the routes to be travelled to achieve these skills will lie through capital transfers from the North. It is principally by means of such transfers that the USA, Canada and Australia reached their present levels of development. And it is precisely these transfers that the Soviets, and more especially the Chinese, have been seeking in order to improve both their agricultural and their industrial capacity.

Attention has been repeatedly drawn to the fact that Western aid is far from disinterested.[10] The instability of commodity markets considerably undermines the Third World. But the effect is not exactly helpful to creditor states either. When commodity prices collapse, so does the ability of the Third World to finance further purchases from developed states. After 1958, increased aid to Third World debtor states, through the IMF and the International Development Association (an affiliate of the World Bank) came about, as Payer has observed, because 'Western policy-makers perceived that expanded aid must supplement contracting trade as a means of financing their exports' (p. 30). Payer vividly describes it as 'debt slavery on an international scale'. Her point is that the IMF does not attempt to assist Third World nations in the essential business of standing on their own feet economically, but rather 'coaches them on how to qualify for increased quantities of new credit' – thus undermining what should be the chief objective (p. 46). The indictment is perfectly valid. It is one thing to claim, with Payer, that 'the nation which wishes to break out of imperialism's grip must not only say nay to the IMF's demands, but must also have the courage to discipline its own consumption and channel it along the most constructive lines' (p. 210) – all of this is perfectly reasonable, if insufficient. It is another matter altogether to claim that, whatever the source, whether East or West, and even without conditions, 'large-scale aid would be a pernicious influence on development', would be merely 'reformist', would worsen

10. See H. Magdoff, *The Age of Imperialism*, New York, 1969; T. Hayter, *Aid as Imperialism*, Harmondsworth, 1971; G. Frank, *Capitalism and Underdevelopment in Latin America*, Harmondsworth, 1971; and C. Payer, *The Debt Trap: The IMF and the Third World*, Harmondsworth, 1974.

the difficulties of the less developed countries, or that a 'high degree' (how high is that?) 'of self-sufficiency' or autarchy is imperative for the sake of survival alone; that 'economic self-sufficiency' or autarchy is possible for any country with sufficient arable land (p. 212). Whatever can be said in the abstract for the potential of economic autarchy in any land, the role of reconstruction played by the Marshall Plan in Europe after the Second World War makes nonsense of the claim that large-scale aid *must* prove pernicious.

It is true that far greater emphasis than previously should be placed on food production for local consumption in African states. But the political reality in any event will not permit a thoroughgoing autarchy. It is clear that poor, fragile and ethnically diverse governments in the Third World are none too difficult to overthrow. Any radical move to disengage, as by the Pol Pot regime in Kampuchea at the end of the 1970s, creates significant domestic discontent and matching or excessive governmental repression. Even where there is no domestic discontent and repression, there is always the possibility of simple invasion by neighbours who have great and powerful friends (even enemies) to supply them with ample weaponry – as in the case of the Vietnamese invasion of Kampuchea (with the 'aid' of abandoned US material). The movement of the Libyan army into northern Chad, the Somali army into Ethiopian-controlled Ogaden, the Tanzanian army into Uganda, several raids by the Smith regime into Zambia and Botswana, South African military threats against Zambia and other African states, together with South African armed attacks upon Angola, Botswana and Mozambique, all illustrate the possibility of invasion by stronger neighbours.

One of the most perilous courses which any contemporary state can attempt to follow consists in putatively turning its back on the rest of the world – even economically. States may well wish to be left alone, but it is a wish which will virtually never be met. Afghanistan was squashed by the Soviets, Tibet by the Chinese, Timor by the Indonesians, Grenada by the USA. The Chinese themselves tried autarchy under Mao Tse-tung. The reversal of this policy, lent such dramatic emphasis by Premier Deng Xiaoping, in the fulsome welcome accorded to President Nixon in 1973, was not the result of whimsy. The future of Hong Kong, Macao, even Taiwan, is not much open to doubt. The Soviet Union was developing apace, as it still is. With Soviet military capacity vastly expanded, China was militarily engaged on her northern marches. The point was reached where political isolation and economic autarchy, creating a vacuum for 'cultural revolution', simply lost its credibility. The prospect of achieving agricultural and industrial development merely by means of self-help ceased to be plausible.

It is one thing to argue for new terms – as with the World Bank – or even for an overhaul of basic structures. It is another matter altogether to argue for simple disengagement. The ultimate consideration is that such an option is no longer (if ever it was) available. A system powerful enough in this age to disengage from the rest of the world is already one secure enough not to need or wish to. The poor are those who are most exploited: they are available and vulnerable. They are also those upon whom the rich bestow alms, when the spirit moves. Being poor, they may expect some charity. Not being fools, they will not expect charity in turn to make them rich (or even necessarily to satisfy their basic needs). The poor cannot ignore the rich – or in this sense 'disengage'. Even when the rich 'ignore' the poor, they still continue to determine their conditions of life.

A British colonial government – with the best motives in the world – may establish a game park and, as a part of this, prohibit the Ik people of Uganda from pursuing their traditional *métier* as hunters. The colonial power may do this in perfect ignorance of such a measure's consequences for the Ik – indeed, without even knowing the name of this people, mistakenly calling them the 'Teuso' or 'Teso'. The colonial policy may even be embraced by the independent, African successor government. The effect upon the Ik, however, was that their culture was grossly impoverished, and also their humanity, and that they have languished and starved.[11]

What happens domestically repeats itself in various ways internationally. When the northern states are hit by recession and in consequence reduce proportionately their expenditure on many of the less essential consumer items they traditionally purchase from Third World states (such as cocoa or cashew nuts or vanilla) it is the poorest and most vulnerable of the mono-economic Third World systems (such as Ghana, Mozambique and Madagascar) which will be worst affected. The massive fall or fluctuation in prices makes the planning of even the most basic and essential services well-nigh impossible – with immediate negative

11. C. Turnbull, *The Mountain People*, London, 1973.

consequences for nutrition, education, the renewal of assets, the maintenance of public order and the like.

The poor or vulnerable are controlled not only indirectly, as suggested above, but also directly – as in the African past through extensive slaving (*c.* 1500–1880) to service New World sugar and cotton plantations. The later colonial subjugation *in situ* was anticipated by the establishment of coastal outposts, the frog-marched recruitment of local and coastal allies, ultimately followed by the massive imposition of direct control over the hinterland (*c.* 1880–1960). In all of this there was involved the overturning of traditional values and the simultaneous installation of entirely new economic systems tied in a subsidiary and subordinate manner (mineral and agricultural) to metropolitan states (as in the extraction of Zaire's ivory and rubber, of Nigeria's palm oil, of Zambia's copper, of Azania's gold and diamonds). The Beauty of independence neither did nor could bring about the automatic metamorphosis of the Beast of longstanding politico-economic arrangements.

There are those nationalists in Africa who will claim that contemporary evils are all to be explained by reference to this past. There are those in the West – on the left and right – who will claim that the remedy lies almost entirely with the Africans themselves, in the domain of self-help. (There are, for example, suggestions of a common position along these lines in Payer as also in the Australian Harries Report.[12]) In the first case, to attribute all evil to external factors is largely to deny the prospect of any future local control over development (of whatever kind). On the other hand, to deny the significant disintegrative effect of external impingement upon Africa would appear to imply an indifference to the continent's plight, a callous abdication of all responsibility, conjoined popularly with a conveniently simple ethnocentric 'explanation' for the condition of gross disadvantage endured.

Whatever overviews may be entertained on the explanatory level regarding the disparities between rich and poor, it will be clear that the former are not content to ignore the latter. The evidence, if anything, is that they are plainly determined not to. It has been observed for example that the success of the CIA and other such Western undercover organizations has been almost exclusively against the Third World states (Guatemala in 1954; Allende's Chile in 1973; Mohammed Mossadegh

---

12. Payer, op. cit.; *Australia and the Third World: Report of the Committee on Australia's Relations with the Third World*, Canberra, 1979.

in Iran in 1953; Zaire in 1960–64; and possibly Nicaragua in the 1980s), not against the Soviet Union and the Eastern bloc generally.[13] Victor Marchetti and John Marks put it in these terms: 'When a President has perceived American interests to be threatened in some faraway land, he has usually been willing to try to change the course of events by sending in the CIA.'[14] Naturally enough, a parallel case against the Soviet KGB has been made by a great number of figures ranging from John Foster Dulles and Allen Dulles up to Margaret Thatcher.[15]

The concern here is not to apportion degrees of involvement – evil, banal or beneficial – but only again to affirm the irreversibility of the involvement of Africa with both First and Second Worlds, whatever form (among many) this involvement may assume. The so-called 'discovery' of Africa was synonymous with her conquest. The 'discovery', 'opening-up' or whatever other euphemisms are employed mean merely that certain more powerful parts of the world became for the first time more fully aware of and caught up in Africa's existence. Africa will not magically become 'undiscovered' or 'uninvolved'. She will not magically cease to be a focus of foreign intrigue and rivalry. From the time of the Portuguese destruction of the Kingdom of the Kongo in the fifteenth century, Africa was hauled willy-nilly into a global political system dominated by Western powers – and she has basically remained so, with slight modifications of form and style, ever since. Independence had not the effect of getting Africa out of the system, but rather – as suggested earlier – of partly altering the terms on which Africa exists within it. Attempts must be made, both in the north and the south, to revise the terms of this involvement. But there is no way of escaping the involvement *per se*.

It has been argued that the process of 'development', even where it seeks to overcome severe international inequalities, proves in turn to be brutal and oppressive. A great deal of interest has been focused recently on the inequalities internal to the developing states. Land values have risen markedly in Lagos, Nairobi, Abidjan and in so many other African capitals. Politicians and civil servants in particular have been well-placed to acquire such land, to attach ever-accelerating rents to it,

13. See P. Agee, *Inside the Company: CIA Diary*, Harmondsworth, 1975; N. Sheehan et al., *The Pentagon Papers*, New York, 1971.

14. V. Marchetti and J. Marks, *The CIA and the Cult of Intelligence*, New York, 1975, p. 349.

15. See J. Barron, *KGB: The Secret Work of Soviet Secret Agents*, New York, 1974.

extending their holdings also in the agricultural sector, engaging in black market activities, exporting capital necessary for local development, detonating consumer demand through conspicuous consumption, and in a number of similar and associated ways increasing the domestic gap between rich and poor. These events have obvious destabilizing implications. And it has been observed that these new elites have developed in intimate association with foreign capital, seeking to exploit local markets in the manner most profitable to them, if not in a manner always consistent with the overall and immediate needs of the host country.

It is essential to try to avoid these pitfalls. But homogenization is itself a pitfall. Any form of specialization – because of the differential skills it presupposes – has a degree of inequality built into it. For a country to be free of foreign rule, it must itself develop trained cadres with appropriate skills. But means are never sufficient for everyone to be so trained. Nor would it be in every citizen's interest for the attempt to be made. To have all poets and no farmers is an untenable politico-economic arrangement. We may dispute why it is that some farmers become poets and the rest stay put. Rival disputants may be elitist or egalitarian, in respect to their advocacy, whether they offer in explanation genetic, psychological, accidental or other factors. But what is certain is (a) that everyone would be worse off if no farmers remained and (b) that the poets, especially where they become rulers (as did former President Senghor and the late President Neto), necessarily come to occupy positions of advantage which enable them, and up to a point indeed require, that they enjoy – as compared to the ordinary citizen – both distinctive and more costly life-styles.

Brutality and exploitation are always to be avoided. But we know of no state in which the attempt has been completely successful. The rest is a matter of priorities. The problem is to assign them. It is sometimes suggested that help for African countries (in whatever form – as aid or trade) is irrelevent in the degree that internal inequity is not overcome.[16] There is some truth in this. But not much. It is obvious that some individuals, even classes of individuals, in African states are vastly wealthy and that further growth will in some degree enhance that wealth. But it is also obvious that the same phenomenon holds for northern

16. S. George, *How the Other Half Dies: The Real Reasons for World Hunger*, Harmondsworth, 1976.

industrial states, without it necessarily being assumed that maldistribution is itself a sufficient argument against any new investment there. (The fortune of the Dallas-based Hunt brothers at the height of the speculative fever in 1979–80 was reckoned to come to about $7.5 billions in silver alone.) There is domestic inequity within all states. But the relevant consideration is the contrast between the Third World and the other two. For example, in Djibouti, in 1985, the unemployment figure was variably set, but by some officials has been put as high as an unpublished 80 per cent. This is perhaps extreme. But, on average, 57·5 per cent of the 'active population' (15–64 years) will be unemployed or underemployed in any given African country. Following the latest reliable figures (1982), 91 per cent of Africans currently live in absolute poverty.

Industrial states with unemployment rates even as high as 8 per cent or more – given unemployment benefits, free and universal secondary education, efficient public health systems, and average per capita incomes in excess of U S $9,000 – are not to be compared with the bulk of Third World populations, which are in an entirely different position. It has been calculated that 'in a typical developing country ... the poorest 40 per cent get only 12 per cent of the income', while in Western countries the poorest 40 per cent get '16 per cent of total income' (World Bank figures).[17] Thus, on these figures, the domestic gap between rich and poor within developing states is on average no more than 4 per cent greater than within developed states. Basically, it is the absolute magnitude of goods and services available for distribution in the rich states which accounts for the vastly higher average standard of living in these countries – not a more even distribution of wealth. Egalitarian distribution is not in fact significantly greater in First World states than in Third World states.

In sum, there can be no doubt about the fact of gross domestic inequality within most Third World states. But inequity, abuse, error, extravagance, waste and more are not restricted to any one set of systems. Certainly, the causes of the cumulative disadvantage characterizing the Third World cannot simply be reduced to such inequalities. Nor can the elimination of all these be regarded as a proper and reasonable precondition for various forms of assistance to African states. Nor

17. See P. Harrison, *Inside the Third World: The Anatomy of Poverty*, Harmondsworth, 1979, p. 414.

should African leaders, usually soldiers, men saddled upon the mute and heavy discontent of the African masses, be encouraged simplistically to assume that through what are merely brutal and summary executions of former leaders – as in Ghana, Nigeria, Liberia and elsewhere – they will achieve anything more than a discharge of pent-up emotions. Of course, such displays of intolerance do achieve marginally more than this: they frighten off both domestic and foreign elites, as well as domestic and foreign capital. Observers who decry Third World economic inequalities, and put forward the elimination of these as the top priority, do not (paradoxically) hesitate to condemn outright the excesses of the earnest and ill-instructed military recruits to their point of view who are swift to fire upon the elite targets which have been marked up and staked out for elimination – as by the European Left, despising all privilege, and sometimes inclined towards terrorism as a tactic, with philosophical anarchism as the goal.

Development – meaning the achievement of greater equality between nations – will most certainly not end all inequity, brutality, exploitation and the rest. But there is no alternative to development. It could be said, naturally enough, that there *is* an alternative. But if this is said, the intended solution must signal a return to subsistence economies. Not every system, in fact virtually none, will choose to attempt that sort of reversal. And those that do will simply be collectively weakened *vis-à-vis* those that do not. Even Gandhi's proposals for village industry and the like, while based upon what he perceived to be traditional Indian values, were designed to make the best use of the limited resources available to achieve an appropriately modern basis for a new collective dignity. Nyerere's *Ujamaa* programme, whether well- or ill-conceived, was similarly designed to move the system forward, to overcome a frightening international imbalance, but with reference and deference again to what were perceived as traditional values, and within the severe limits of exiguous resources. Limited resources impose a constraint; they do not provide an excuse for attempting to opt out.

On 26 April 1965, Julius Nyerere spoke of a lesson he thought could be learnt from China. He put it thus: 'It would be very foolish for us to borrow money from other countries to do things which we could do for ourselves; we have already experienced the danger of doing this. And it would be very foolish of us to let our Development Plan fail because we want ... to appear as rich as a country like America. Some of the things we have done in the past, like buying big cars for the Regional

Commissioners, were bad mistakes of this kind. They must not be repeated.'[18] Leaders in developing countries need incentives. The citizenry, on the other hand, need morale. Excessive reward for the few is wasteful. But to draw the line between minima and maxima is the problem. Excessive reward for the few is not only wasteful, it is demoralizing to the many. The reasons then for tackling domestic maldistribution, as for accepting a degree of inequality, are complex. A policy which pushes too far in either direction may find itself in serious trouble.

A part of the suggestion caught up in many arguments against domestic African inequality is that – being in essentials externally caused – it can be overcome by turning one's back on the international order. But there also existed domestic inequality in Africa before the European conquest. There was stratification among the Alur, slavery among the Ila, castes among the Rwanda and so on.[19] Granted, following Ronald Cohen, 'there is widespread lack of differentiation ... in traditional African societies'.[20] But there was differentiation none the less, with its inescapable inegalitarian implications. The foreign connection did not generate all of it. Nor would the elimination of that connection eliminate all of it.

If African states could disengage (*per impossibile*), by reverting to subsistence systems, untroubled by foreign intruders, there would still remain sufficient economic and technological means within Africa for local leaders to play the tyrant – as happened during the *Mfecane* of the last century, with its attendant Zulu and other expansionism. We may see from contemporary experience, indeed, that it is often precisely those African governments which seek or achieve least traffic with the outside world that end up displaying the most oppressive behaviour: Uganda under Idi Amin, Equatorial Guinea under Macías Nguema, the Central African Republic under Jean-Bedel Bokassa. For these and similar reasons, any injunction to 'disengage from the international system' has about it an odour of illogic, of double-stink – not only because it cannot be done, but also because, even if it could, it would not solve the problem.

The present international system is itself almost as new as the new African states. The novelty of each is lent emphasis by their mutual involvement or engagement. The rather large notion of 'involvement'

18. J. Nyerere, *Freedom and Unity*, Oxford, 1967, pp. 323–3.
19. A Tuden and L. Plotnicov (eds.), *Social Stratification in Africa*, New York, 1970.
20. In J. N. Paden and E. W. Soja (eds.), *The African Experience*, 2 vols., London, 1970.

does not reduce to the narrower notion of 'aid'. If the grand formula of disengagement merely reduces to the idea of stopping 'aid' then we have a quite specific issue to consider. We used to hear frequently the declaration: 'trade not aid'. And if we ruled out aid, the trade – which already exists – would none the less continue. The involvement, the necessary lack of 'disengagement', would continue. Every schoolboy knows that the chief problem about trade is not that it does not exist, but that it is so arranged that primary producers get little or nothing out of it. Commodity markets are highly unstable, and nowadays, even for crucial items like oil, it is a buyer's market. Northern buyers are able to stockpile, until prices decline to a level which they find attractive. There is far more to be said in all this. And a great deal is being attempted – by the Third World states themselves – to improve terms. Hence the importance of negotiations for a general system of preferences and more generally for a new international economic order. Not that these moves have proved highly successful.

Although aid may be optional – even where engagement is not – it would be irresponsible for most African states to attempt entirely to forgo it, as irresponsible as for wealthier states to refuse. There is involved here, as everywhere in human relations, a question of generosity and good-will. But there is also involved a question of mutual interest. Governments and elites which cannot cater to the basic needs of those they govern are obviously at risk. Instability in the developing world, which is most of the world, cannot afford any comfort to northern states. The north must be more forthcoming. And the south must press its arguments and improve its organization to help secure the intended effect. The only rationale for aid, where it is forthcoming, is that it should genuinely aid. Without Chinese aid the Tanzam railroad would not have been constructed. Without Soviet aid, Lake Nasser would not exist (there would be no Aswan to hold it in). Without the Marshall Plan, virtually all of Europe today would probably remain grossly undernourished, even impoverished, and marked more by war than by occasional bursts of terrorism. Aid donors were no less humanitarian, in these cases, for pursuing their national interests. The Chinese (in East and Central Africa) sought to contain the Soviets; the Soviets, the Americans (in Northern Africa); and the Americans, the Soviets (in Western Europe). There was a perceived net benefit to both donors and recipients. It is not always so, of course, but failure is not an argument for ceasing to try.

If, in the course of commercial transactions, capital is lent to the Third World, it must normally be paid for. But over time, the situation may be reached where repayment on interest becomes so prohibitive, as in a country like Sudan or Zaire or Ghana, that to continue such debt servicing appears to be wholly inconsistent with the wider goal of development. And international organizations, like the IMF, may be largely to blame. To avoid this, states may appeal for aid, whether in the form of rescheduling, a simple moratorium or outright cancellation. But generally the chief object is to facilitate the transfer of technological innovation to the Third World without compounding the economic difficulties that are normally associated with this.

Aid never comes without some 'strings attached', strings of various kinds and strengths. On the most obvious level, there will remain the repayment of interest, even if at reduced rates. Alternatively, aid may be conceded on condition that the recipient utilizes personnel, procedures or products from the donor country. Where aid is multilateral, as from an agency such as the World Bank, the host country must meet the conditions of the donor agency. And those donor states contributing most to the agency may insist upon weighted representation in that agency in order to retain more than equal say. Donors of whatever kind will always require *some* oversight role. Where transfers of technology are sought through ordinary commercial means, recipient states may be met with ruinously high interest rates. By contrast, to accept simple bilateral or multilateral aid always invites a more direct form of oversight role by the external agencies supplying aid. Many observers (like Payer) are suspicious of all aid. But any generalized prohibition would be irresponsible towards Africa's peoples and their hopes for the future.

From this point, we are no longer dealing with the question of whether development as such is desirable or not. Nor, in present circumstances, with whether aid is desirable or not. The ticklish question is how best to develop, and how to direct aid to achieve this effect. One may confuse what is essential with what is not. African mothers in insanitary conditions have in the past been encouraged by unscrupulous producers to switch from breasts to bottles in the nursing of infants – an engagement not only inessential but always dangerous (given unclean water and poor hygiene) to the health of infants so fed. The remedy available to local governments – if not the courage and interest – is obvious. None the less, useless or harmful drugs, luxury cars and Coca-Cola are not in the same category as roads, canals, dams and myriad other forms of technological innovation – all of which, in principle, produce surpluses, which are necessary to provide the basis for a more distinctly human life-style. It would seem reasonable to say, if loosely, that Africa cannot justify expending foreign exchange upon much that it buys. But materials for roads, improved secondary education, etc. do not, by contrast, fall into the same category. Foreign exchange will always require to be expended as part of the push for development.

Between independent states, as between individuals, there exists conflict and competition, cooperation and fellowship. But whatever the type of state, there is a difficulty where any system is dramatically vulnerable *vis-à-vis* its fellows. If we assume a democratic model, the profile of government must somehow replicate the profile of its citizenry. On this model, further, the government must broadly represent the interests of its citizenry, as somehow advised by the citizenry itself (although complete advisement is never possible). In as far as we assume that every government generally attempts to pursue the interests of its own citizenry (just as rival football teams do not generally seek to score against themselves) the gross vulnerability of one government *vis-à-vis* another will constitute a risk for the citizens of the weaker system. A domestic government, in as far as it seeks to satisfy the demands of its own citizens, as well as those advanced by a stronger government or governments acting upon it, must mediate inputs from at least two opposed sources, in the course of which domestic constituencies are almost always likely in some degree to suffer. Many of these domestic constituencies may themselves be mutually opposed. But the point about powerful foreign states is that, should they advance demands, they themselves become *de facto* domestic constituencies, often able to override other domestic actors, retaining only a formal appearance of alien non-interference. Any government seeking to preserve itself, or even to establish itself, and confronted with a threat or a promise of help from a more powerful neighbour, is always to be expected in some degree to bend the knee.

In the degree that a government is vulnerable (most importantly on the military and economic levels) *vis-à-vis* fellow governments, in just such degree is it provided with an inducement to serve the interests of the latter. A government that is compelled to take account of the pressures of agents external to it has a diminished capacity to promote the interests of its own citizenry. A government may well become more domestically authoritarian as a means of strengthening its position against external powers. A government, however, which yields to the pressures of more powerful neighbours (thereby generating domestic discontent) may well become more domestically oppressive as a means of strengthening its position against internal dissidents.

If a more powerful state exploits the vulnerability of one less powerful, and acquires thereby a degree of control over it, the advantages which this control confers can be fed back (by those who govern) into the dominant state in such a way as to diminish the power of ordinary citizens there and to enhance that of rulers. The 'overseas' dependency is basically responsive to the government of the dominant power, not to the citizenry of that power. Because of the reciprocal external support it receives, the dependency can be less responsive to its own citizenry. Finally, a dependency provides bases from which, and thus means by which, the rulers of the dominant state may themselves distort such a state's democratic profile (in the degree that it exists). Continued French control in Algeria, Portuguese control in Africa, American control in Viet-Nam all seriously threatened or inhibited democratic rule at home. The dependent state represents an external base whence dominant metropolitan elites may, for example, recycle mercenaries and assassins and launder money – so as to give unfair advantage to certain metropolitan candidates up for election or in order to promote certain views without identifying the proponents.

In sum, gross inequalities between states generate increased repression within the weaker systems, subsequently externalize important sources of support for the stronger states, and provide a means by which the democratic character of the more powerful states may themselves be further undermined. All such processes can only deepen international tensions, especially between the more important contenders for international power – not only at the expense of the citizenry of the poorest states, but ultimately also at the expense of the citizenry of the richest. Such developments threaten the globe as a whole. The move by the poor towards equalization – essentially through multilateral organization on both regional and global levels – must obviously serve their interests, while also serving, if in a manner less immediately evident, the interests of humankind. The logic of development, however distorted its manifestations, is also to be seen as engaging the logic of survival. Survival of course is not the *only* value, nor one that is invariably appropriate. It is not, all the same, the least of values, nor one to be lightly spurned.

To survive, people may require to act in ways they would prefer – at least in the abstract – to avoid. They may adopt abhorrent measures with a view, simply, to improving survival prospects. One of the gauntlets commonly run is self-brutalization, resorted to so as to sidestep brutalization at the hands of others. Many of those detained at the pleasure of the 'State Research Bureau', at Makindye prison and elsewhere under the Amin regime in Uganda (1971–9), were compelled to hammer to death fellow inmates, under pain of being themselves summarily dispatched. Better, really, to die than brutalize oneself in such a degree. But most self-brutalization is rarely so extreme as that, or so sudden. One of the most important and brutal transitions in the history of the human species is from a condition of hunter–gatherer, or nomadic, or settled agrarian intimacy, with a great emphasis upon friendship, to modern commercial and more especially industrial arrangements, with the emphasis upon efficiency. (It is sometimes referred to as the shift from *Gemeinschaft* to *Gesellschaft*, following Ferdinand Tönnies.) We have adjusted so gradually to the change that we are commonly disposed to disdain friendship societies as 'primitive', while celebrating our more antiseptic accommodations as 'civilized'.

If we put to one side most of the praise (mere ethnocentrism really) that is associated with 'civilization', we can spot something else at work beneath the hype, something fairly objective, and this is the fact that the 'civilized' are politically and economically on top of the 'primitive'. The primitive is merely the country bumpkin in the midst of city slickers, the sucker (adapting Barnum) who is born every minute into a Machiavellian global village. She or he is the Eskimo (Inuit) of Alaska or Canada, the Amerindian of the US West or of the Brazilian Amazon, the Zulu in Johannesburg, the Masai in Nairobi, the Aboriginal in Australia, the South in the North. The taste of effective disengagement is sweet. The trouble is that other folk cannot, will not, *laisser faire*. The closer a people approach to the condition of *being* (not abstractedly *wishing* to be) disengaged, the easier in general does it become for other states to snatch the sweet away. A small independent community of limited technology occupies space only in a sort of state of nature, where it is unaware of the rest of the world and vice versa. But once the sacred enclosure is shattered, what was independent becomes subject, and 'the state of nature' is transformed into 'primitivism'.

The treatment handed out to the survivors from the state of nature is rarely encouraging. The once-proud pastoralist of the Sahel, or of the Danakil, or of the Ogaden, forced from less to more marginal pasture, eventually becomes a derelict in the cities or in the camps. The Aborigines of Australia, expelled early on from the land, have by now clearly failed in the bid for 'land rights'. Their numbers lie at less than 1 per cent of the total, while among them circulatory deaths exceed by twenty times those of whites, maternal deaths by ten times, trachoma and other diseases betraying similarly exaggerated incidences. Aborigines are ready victims of police brutality and in Western Australia comprise 33 per cent of total prison population. One well-placed Australian doctor (Professor Fred Hollows) claimed these people to be 'destined to die out quite rapidly' (August 1985). The demoralized visage of the contemporary Amerindian is not much more encouraging. A powerful, affluent, aggressive America has made an approximately clean sweep of the native American, whose numbers are much less than 1 per cent of the total, constituting a variety of largely illiterate and broken folk who bear mute testimony to what the most defenceless peoples are heir to in consequence of total defeat – especially where the phenotype itself records the loss. People may – it is not the case that they always do – place the means of survival at par with the evil they would suffer in going under. To keep abreast may involve becoming something worse than one was, or thought oneself to be, in some earlier age, in some more pristine state. But not to keep abreast may involve being transformed into something still worse, at the hands of others who have acquired the means.

The process of development – with its accelerated movement away from subsistence structures and towards increased professional specialization, with widening power gaps between components of the citizenry and comon differential pyramiding of income and status indices, and with associated forms of military, economic and other subjection to external interests – may be conceived as rendering the achievement of democracy impossible. The very competition involved in the process often renders rulers insensitive to the genuine needs and actual hopes of their supposed constituents. But as against all this, our chief consideration must be that those states which most earnestly attempt to stand aside from the scramble will also be those which must grow increasingly vulnerable to erstwhile competitors who end up atop the heap. Those left at the bottom will almost certainly become more rather than less authoritarian. The colonial regimes were by definition undemocratic and arose out of a marginally or distinctly superior technology permitting highly disciplined – and autocratic – control over the colonized. The pace of technological innovation is not decelerating – quite the contrary. Colonization, paradoxically (and in this sense beneficially), reduced the technological gap between the colonizers and their potential indigenous replacements.

This shrinking of the technological gap, however minimal, was to achieve significant liberating effects as so dramatically demonstrated in the success of the Algerian war of independence (1954–62), in Angola, Guinea-Bissau and Mozambique (1961–74) and most recently in Zimbabwe (1965–80). The existence of a countervailing, and sometimes materially supportive, Soviet power was clearly crucial. It became clear that the defeat of a colonial power, like Portugal, could provide an indispensable basis for the democratization even of the colonizing state itself. In this sense, Africans and Portuguese would have ample grounds for viewing the military eclipse of the latter as an ultimate moral victory for both sides. On the economic front, African states will have to press forward in earnest – just to stand still. To surrender simplistic notions of a golden past leaves little room for contemplation of a hallowed present. No contemporary African state, not even Ethiopia, is traditional. All are far more highly centralized, even now, than ever were their counterparts in the past. It cannot be reasonable to suppose that future African states, once they have become technologically more sophisticated and thus better able to register and meet the basic needs of their citizenry, are condemned to prove *less* democratic than those governments by whom the mass of African peoples are presently governed.

It is undeniable that the process of development constitutes a pursuit of power. Also, a reasonable pursuit may well degenerate into a vulgar obsession. All the same, to try to become more powerful is not the same thing as glorifying power, although exercised in various ways and degrees. What one cannot meaningfully do is to ignore it, or to assume that one can ignore it. It is reasonable for governments to seek to obtain sufficient power to place a brake on the prospect of themselves and those they may represent being either imperiously or brutally dealt with. It is unreasonable, in the face of history, to suppose that subjection to such treatment is of low probability in the case of those systems which remain markedly vulnerable to others. All obsessions, including power obsessions, are vulgar. But the determination to retain some control over one's destiny is a concern that is not entirely irrelevant, and is perhaps even sane. It is of course possible for persons and polities so to fuse their wills with the urge to power that they make themselves its perfect slaves. On the other hand, it is everywhere recognized that it is only the powerless who are in fact made slaves. As against all this, it still remains possible for power, in some acceptable sense, to make men *free*: to give them collectively greater control over their environment, both human and natural, than circumstances would otherwise permit. What certainly should be recognized is that a ceaseless striving for power after power, on an unrestricted individual or collective basis, is likely to yield – as it so often has – nothing more than domestic or international chaos. A workable collective security system ultimately serves the best interests of both the weak and the strong. Even the strong, as now in the case of the United States, do not always remain so. The medically insured – assuming sufficient numbers of these to render individual subscriptions a less than onerous burden – have no cause to lament the prudence they display when, being young and strong and least in need, they pay for cover which is likely, with advancing age, to serve them as a boon.

The basic thrust of development then is to be understood as engaging a power pursuit of an equalizing kind. The pursuit will also involve vagaries, idiocy, abuse. But to deny it would prove plainly, on balance, both inhumane and irresponsible. It is tempting to seek to limit the reach of this engagement by attempting to stipulate a minimum of 'basic needs'.[21] This is not so great a problem – fixing basic needs – when these needs are utterly basic, as for example in a situation of mass starvation. On the level of nutrition, we might stipulate as a healthy minimum the intake of 2,500 calories per day. The problem is that to suggest anything beyond this is necessarily to desert any firm concept of basic needs, and to enter into a troublesome desert of comparison. The sturdy Volkswagen in nineteenth-century Germany would most certainly have been regarded as a remarkable invention and an incomparable luxury. But of course in the contemporary Bundesrepublik it can only be perceived as a rather mean and stingy piece of engineering. It is a mistake to suppose that we can escape this sort of fact – along with the competitive, comparative assessment that accompanies it – except in a frozen sort of setting where there is nothing to compare. Paradoxically, a state would have to ascend the summit of technological achievement just to create the possibility of blotting out an envious awareness of the

21. E. Fromm, *Man for Himself*, New York, 1947; A. Maslow, *Motivation and Personality*, New York, 1954; H. Marcuse, *One Dimensional Man*, London, 1954; C. Bay, *Structure of Freedom*, Stanford, 1970.

rest of the world – thus to keep the achievements of other peoples and states from tweaking the nose of its own people's self-esteem in various spheres. But states so well-placed are, of course, more committed to making themselves known than obscure. Their vocation often assumes an approximate or actual religious aura – whether we have in mind capitalist America, communist Russia or the Islamic revolutionaries of Libya. In short, not being isolated nor being capable of isolation from other peoples, styles and achievements, the developing African states can scarcely presume to fix any basic needs beyond those most obviously essential to survival. The perception of what is needed will always be revised upwards as soon as the initial minima set are reached. The idea of being rationally able to fix basic needs in the abstract and for all time, even if not a complete non-starter, would not appear likely to carry us very far.

There is no way in which we can provide 'adequate material conditions' *per se*. It is as well to recognize this clearly. If it were possible for an individual or a group to become totally isolated from the rest of the human environment in which we are all immersed it might perhaps be possible to entertain the notion that needs are not relative – at the very least there would be nothing to relate them to. But this is not our situation. Nor is it more generally a characteristic of the human condition. Whether we are competing or not, we cannot remain immune to the fact, say, that the hunter in the next field employs a shotgun while we make do with a boomerang – or a blowgun. At the very least it is likely that he will bag his birds and scare ours away. If time and efficiency are important, and if the birds do stay on, the shotgun will certainly command greater leisure time for its owner, and should he choose to avail himself of it, he will be freed to cultivate other aspects of his being. Should the hunter with the shotgun advise the neighbouring hunter, who is without, that 'the difference makes no difference', and that 'men with blowguns only must live fuller lives', the sincerity and credibility of the one in the eyes of the other might prove subject to deflationary pressures. Those who wing about the world in jumbo jets are not well placed to suggest to others who do not that the latter have no need of these things. Those, alas, who argue so would do best to surrender their aeroplanes, radios, telephones, cables, photocopying machines, projectors, electric typewriters and Levi-Strauss jeans before they deign to speak – or preach – about the simple things in life being best. The poor of this world, like the rich, do not, for better or worse, occupy an island state. Parisian innovations are not delayed over-long in filtering through to Abidjan. They have an impact upon human awareness and must revise, often radically, the local awareness of what is 'needed'.

Sometimes it is thought that, if states can fix upon basic needs, and aim for these, it will prove possible and easy to achieve various forms of national and sub-national autonomy. The African independence movement could be seen as having involved such a quest, as observed before. But, there is no prospect for such autonomy. The independence movement itself certainly did not lead that way. A useful distinction is to be observed between independence and autonomy. To be autonomous is to be self-sufficiently alone. To be independent, by contrast, is merely to transfer politico-legal sovereignty from one locus to another. It has nothing necessarily to do with autonomy, self-sufficiency, autarchy. If we ask what independence has brought about, and whether it has cut Africa off from the rest of the world, the answer is firmly negative. What independence has achieved is precisely the opposite – tying Africa more closely than ever before to the rest of the world – at the UN, in diplomatic missions, through commercial transactions, student exchanges and in many other ways. Whatever else independence may have brought, it is not autonomy. And judging from what independence has brought – not accidentally but deliberately – it would seem that autonomy was never really envisaged.

In the end, there is no substitute for judgement. In the Third World, as in the West, there can be no question but that an appetite is artificially whetted for a great variety of non-essentials. But the distinction between mere appetite, on the one hand, and genuine need on the other, while clear in the abstract, is again less clear in practice. The mammoth groundnut scheme initiated by the British in Tanganyika after the Second World War at a cost of £36m. sterling was a complete failure. Ghana's Nkrumah, in this respect, was equally grandiose – expending millions upon a giant reception centre for African heads of state ('Job 600'); securing fragile and utterly inappropriate eastern European tractors to clear and work the dense forests of Ashanti. History is replete with projects which, to those initiating them, appeared sensible enough, but which prove in retrospect to be appetitive pipedreams. By the same token, some of the soundest (if not especially significant) *economic* schemes throughout the continent (tourism in Kenya and Egypt or soft-drink and beer-bottling operations everywhere, for instance) have not been directly associated with any particularly noble aspirations. Again, we draw what we take to be a clean distinction in principle between the essential and the non-essential. But the distinction does not translate out so cogently on the level of practice. On this level, one may as easily be led astray by incremental appetite as by the lure of a grand design. There is no substitute for judgement. In present circumstances, the notion of African development is not merely appropriate, but imperative.

Only the strong can possibly isolate themselves. Being strong, they do not feel the need to do so. The weak cannot isolate themselves. In their exposure, they sometimes complain that they merely wish to be left alone. It is a luxury the rest of the world, unfortunately, will not allow. Africa, whether she likes it or not, forms part of and is an actor within a global system – economically, politically and culturally. Her isolation has been steadily eroded since the fifteenth century, quite beyond all prospect of reversal. Independence was merely a step towards redefining the African role within this evolving system. It in no way provided an opportunity to opt out of it. Were one able to opt out of the system, one would not be labelled as 'backward', 'underdeveloped', 'developing' or 'modernizing' within it. Of course, the 'system' is not frozen. It is refractory enough all the same. There is no choice but to make every possible exertion to change it. In the course of doing so, it is as well to jettison that unreconstructed fear of power which so commonly assails persons of conscience. It is an indulgence as irrelevant as it is unavailing. African peoples have stared power in the face far too long to be unnerved by it. It has worked like a lash upon the continent, uprooting entire societies, destroying local structures, generating war upon war, tearing the people from the land, transporting them in an agony across the sea, forcing the development of a new world through the instrumentality of unrequited labour, finally achieving the direct subjugation of Africa itself through colonial conquest, whether by the ruthless tactics of Bugeaud against the Algerians or the ghastly atrocities of Leopold and Stanley against the Congolese or the savage 'extermination order' of von Trotha which shredded tens of thousands of the Herero. To fear power is to be immobilized by it. From there is but a short step to becoming its victim. The chief – not only – defence against it is participation in its exercise. What is to be feared, and opposed, is not power, but unfair, unjust and grossly unequal use of it.

There can be no question of Africa ever being able to repay to Europe the dubious favours extended over several hundred years. Nor should there be. The world is full of grievances, and legitimate grievances, but to become fixated upon these is more debilitating than to grow large in the light of prospective reconciliation. To forgive, however, is not to forget. For those who forget history, as has been observed, are condemned to repeat it. To avoid such repetition is vital for Africa, for the Third World, but also for those states which today are globally dominant. The Americans and the Soviets are obsessed by the problem of security. But their citizenry rarely stop to think that most of the world's peoples have no security at all, not even social security, and obviously no military security, against these very giants who strain none the less to defend themselves. If they do not come to blows first, and in the process destroy or further cripple us all, perhaps it will only be because the African states, in the company of other Third World associates (not without 'aid' from smaller 'developed' states), refused to surrender at least the tattered remnants of political neutrality, of non-alignment, of independent conscience and judgement in the face of a potentially unrestrained rumble between capitalized and socialized barbarisms. The looming struggle is only *apparently* between *them*. But should it supervene, it would quickly envelop everyone. The fewer recruits the protagonists co-opt, the less dangerously polarized will the world be. One problem is to get the strong to recognize their own ultimate interest in making the powerless stronger, if only to reinforce independent judgement in respect to their own otherwise dangerous differences. To labour for development – which implies more equal strenghth and improved well-being – is also to labour for the continued survival of humanity.

# Epilogue

The pictures of horror depicting the torment of so many African children, pictures which flooded TV screens and the press at Christmas 1984, had largely faded from view by Christmas 1985. The misery had not gone away, but its volume had for a time subsided. It was less visible. If the wound was deep, one could not so easily detect how deep a wound it was.

Africa's peoples were being copiously bled; dramatically here, silently there, but emphatically everywhere. It is not of course the case, where a limb is gashed, that the entire body is mutilated; not so, where elements of a nation are utterly ruined, that all of that nation are destitute; not so, where some regions of a state are churned to dust, that the state as a whole betrays no life. Life goes on, especially when retrieved from the brink of disaster. But it is a demeaned life: crowded, displaced, indebted, cramped, disputed – and envenomed by war.

Drought in Africa goes back a long way. Its effects are the more acute the lower the reserves available to the people who have to fight it. Parts of the Horn of Africa were waterless for four years, parts of Sudan for twelve, Cape Verde for eighteen. The longer the drought, the more devastating its effects. But the problem is not just one of duration, but of this in relation to available reserves. Because Africa's resources are very limited, the consequences of drought are the more terrible. All developments or structures or relationships which further deplete or deflect Africa's resources have the cumulative effect of diminishing, and indeed of extirpating, life – human and animal.

In 1986, the scale of Africa's famine had declined. But the projected population for the continent by the year 2,000 was 800 million. The population is growing at a faster rate than is food production. In fact, food production per capita has declined in the past twenty-two years by as much as 25 per cent. Africa is growing more and more for export, but less and less for consumption. Since the cost of buying food from abroad has vastly expanded, while the return on exported African cash crops has plummeted, there has opened up a chasm between the need for food and its availability. The image of misery projected by Africa upon the world scene will wax and wane. But there is no prospect of it simply fading away. The structural difficulties are enormous. And it will require more than a passing revulsion to put an effective end to such gross suffering, whether or not it is (usually it is not) seen.

People need not survive in the way that is increasingly characteristic of African conditions. They need not do so because there are resources in the wider world which can be brought to bear to bind these wounds, and because it is possible for present arrangements to be amended. The world is an integrated whole, for good or ill. It is a question of people accepting their responsibilities, on whatever level these happen to fall, and acting accordingly. The first and most important task is to prevent death by starvation. But the best way to do this, to prevent its recurrence, is by eliminating the conditions which lead to it. And in order to remove these conditions, it is important to recognize them for what they are.

African states have their share of responsibility for the deaths that African children, especially, have had to face. But the responsibility is far from being theirs alone. African states have been shaped by historical forces which gathered first elsewhere and struck last in Africa. The Brandt Commission reports anticipated the difficulties in which Africa was to find itself. In May and June 1983, the Director General of the UN Food and Agriculture Organization was to do the same. But no serious heed was paid to these warnings, or any concerted effort made to avert the impending disaster. African mistakes compounded an existing difficulty; they did not create it. Africa has been pulled willy-nilly into a global system that is not, essentially, of its own making. And it has suffered severely very much because of its position within this system.

Africa has had imposed upon it a role it cannot wilfully escape: as a supplier of primary commodities. It has been compelled to exchange these against the secondary commodities (and related services) of the industrial world. But it is in the industrial world that the terms of this exchange are set. The value of primary commodities (oil apart) declined over 1980–82 alone by as much as 27 per cent. The value of industrial commodities simultaneously rose. The gap between these values, where primary and secondary commodities are exchanged, is plastered over by credit, from north and south, and thus debt. In 1986, the debt of the Third World as a whole was approaching US $1,000 billion. It was well over $400m. for Latin America, and about $170b. for Africa. Africa's

situation, as it was more fragile, produced more devastating consequences. Thus almost $13b. will be spent over 1985–7 servicing the
long-term debt alone (accumulated by Africa for the period ending in
1982). The real interest rate paid on Africa's debt is likely to be over 20
per cent. This means that virtually all Africa's declining export earnings
will increasingly be overtaken by this debt. Countries so burdened have
a markedly reduced ability to attend to costly crises, whatever their
origin and nature. As Maurice Strong has argued, the debt should be
written off, not just re-scheduled.

Africa can perfectly well feed herself. But to achieve this requires
long-term credits for fertilizers, insecticides, roads, communications,
improved machinery and techniques at less than disembowelling interest
rates. Aid for the Third World (as a percentage of GNP) has not been
going up, as ought to have been the case, but down (from 0·51 per cent
in 1960 to perhaps 0·31 per cent in 1986). It has not only gone down,
but has moved increasingly from a more rational multilateral basis (US
$300m. less in 1984 *vis-à-vis* 1983) to a more primitive and self-interested
bilateral basis. The concern promoted from various quarters to develop
a New International Economic Order has been all but ignored. The
United Nations system itself is increasingly by-passed, even spiked,
while being consistently attacked for not doing what many of its more
powerful, developed members will not allow it to do.

The most important of the needs of the Third World is development.
And development is the least likely outcome in time of war, civil or
other. War is almost a part of the structure of modern African states.
And Africa must be helped to overcome the difficulty. The peoples of
the Sahel, for example, did not ask to be occupied by France, to be
forced (by taxes on individuals, huts, cattle and land) to cultivate
cotton and peanuts to pay for an unwanted French administration. But
occupation is what they got, associated with the new cash crops for
export, developed on the new plantations, which accelerated the exhaustion of the land, which crowded out acreage for the sorghum and millet
normally consumed by locals, which restricted the land available to
pastoralists, and which has largely destroyed the entire region under the
formal aegis of the new indigenous regimes which came to power in the
early sixties.

The Sahel needs help, not civil wars. The Sahel needs help, not desert
wars, as that between Burkina Faso and Mali, an ideologically inspired
conflict encouraged by external powers. The Sahel needs help, not quiet,

brutish exploitation which drives people into the sand. If Africa gets help, it will not be because of, but in spite of, Western governments. It will be because of popular awareness of the interdependence between North and South, awareness of a common responsibility of each to the other. Jaded and self-indulgent governmental bureaucracies, it would appear, move best to the rhythm of the hot pin. This primitive form of stimulation is seldom supplied by public concern. But the popular response to the image of a starving Ethiopia in and after October 1984 was of some such kind. It is an example which we may hope will never grow cold.

# Index

## MORE ABOUT PENGUINS, PELICANS, PEREGRINES AND PUFFINS

For further information about books available from Penguins please write to Dept EP, Penguin Books Ltd, Harmondsworth, Middlesex UB7 ODA.

*In the U.S.A.*: For a complete list of books available from Penguins in the United States write to Dept DG, Penguin Books, 299 Murray Hill Parkway, East Rutherford, New Jersey 07073.

*In Canada*: For a complete list of books available from Penguins in Canada write to Penguin Books Canada Ltd, 2801 John Street, Markham, Ontario L3R 1B4.

*In Australia*: For a complete list of books available from Penguins in Australia write to the Marketing Department, Penguin Books Australia Ltd, P.O. Box 257, Ringwood, Victoria 3134.

*In New Zealand*: For a complete list of books available from Penguins in New Zealand write to the Marketing Department, Penguin Books (N.Z.) Ltd, Private Bag, Takapuna, Auckland 9.

*In India*: For a complete list of books available from Penguins in India write to Penguin Overseas Ltd, 706 Eros Apartments, 56 Nehru Place, New Delhi 110019.

# A CHOICE OF PENGUINS

☐ *The Complete Penguin Stereo Record and Cassette Guide*
**Greenfield, Layton and March**                    £7.95

A new edition, now including information on compact discs. 'One of the few indispensables on the record collector's bookshelf' – *Gramophone*

☐ *Selected Letters of Malcolm Lowry*
**Edited by Harvey Breit and Margerie Bonner Lowry** £5.95

'Lowry emerges from these letters not only as an extremely interesting man, but also a lovable one' – Philip Toynbee

☐ *The First Day on the Somme*
**Martin Middlebrook**                              £3.95

1 July 1916 was the blackest day of slaughter in the history of the British Army. 'The soldiers receive the best service a historian can provide: their story told in their own words' – *Guardian*

☐ *A Better Class of Person* **John Osborne**       £2.50

The playwright's autobiography, 1929–56. 'Splendidly enjoyable' – John Mortimer. 'One of the best, richest and most bitterly truthful autobiographies that I have ever read' – Melvyn Bragg

☐ *The Winning Streak* **Goldsmith and Clutterbuck**  £2.95

Marks & Spencer, Saatchi & Saatchi, United Biscuits, GEC . . . The UK's top companies reveal their formulas for success, in an important and stimulating book that no British manager can afford to ignore.

☐ *The First World War* **A. J. P. Taylor**         £4.95

'He manages in some 200 illustrated pages to say almost everything that is important . . . A special text . . . a remarkable collection of photographs' – *Observer*

# A CHOICE OF PENGUINS

☐ *Man and the Natural World* **Keith Thomas** £4.95

Changing attitudes in England, 1500–1800. 'An encyclopedic study of man's relationship to animals and plants . . . a book to read again and again' – Paul Theroux, *Sunday Times* Books of the Year

☐ *Jean Rhys: Letters 1931–66*
 ·**Edited by Francis Wyndham and Diana Melly** £4.95

'Eloquent and invaluable . . . her life emerges, and with it a portrait of an unexpectedly indomitable figure' – Marina Warner in the *Sunday Times*

☐ *The French Revolution* **Christopher Hibbert** £4.95

'One of the best accounts of the Revolution that I know . . . Mr Hibbert is outstanding' – J. H. Plumb in the *Sunday Telegraph*

☐ *Isak Dinesen* **Judith Thurman** £4.95

The acclaimed life of Karen Blixen, 'beautiful bride, disappointed wife, radiant lover, bereft and widowed woman, writer, sibyl, Scheherazade, child of Lucifer, Baroness; always a unique human being . . . an assiduously researched and finely narrated biography' – *Books & Bookmen*

☐ *The Amateur Naturalist*
 **Gerald Durrell with Lee Durrell** £4.95

'Delight . . . on every page . . . packed with authoritative writing, learning without pomposity . . . it represents a real bargain' – *The Times Educational Supplement*. 'What treats are in store for the average British household' – *Daily Express*

☐ *When the Wind Blows* **Raymond Briggs** £2.95

'A visual parable against nuclear war: all the more chilling for being in the form of a strip cartoon' – *Sunday Times*. 'The most eloquent anti-Bomb statement you are likely to read' – *Daily Mail*

# A CHOICE OF
# PELICANS AND PEREGRINES

☐ **The Knight, the Lady and the Priest**
  **Georges Duby**                                    £6.95

The acclaimed study of the making of modern marriage in medieval France. 'He has traced this story – sometimes amusing, often horrifying, always startling – in a series of brilliant vignettes' – *Observer*

☐ **The Limits of Soviet Power  Jonathan Steele**      £3.95

The Kremlin's foreign policy – Brezhnev to Chernenko, is discussed in this informed, informative 'wholly invaluable and extraordinarily timely study' – *Guardian*

☐ **Understanding Organizations  Charles B. Handy**    £4.95

Third Edition. Designed as a practical source-book for managers, this Pelican looks at the concepts, key issues and current fashions in tackling organizational problems.

☐ **The Pelican Freud Library: Volume 12**            £5.95

Containing the major essays: *Civilization, Society and Religion, Group Psychology* and *Civilization and Its Discontents*, plus other works.

☐ **Windows on the Mind  Erich Harth**                £4.95

Is there a physical explanation for the various phenomena that we call 'mind'? Professor Harth takes in age-old philosophers as well as the latest neuroscientific theories in his masterly study of memory, perception, free will, selfhood, sensation and other richly controversial fields.

☐ **The Pelican History of the World**
  **J. M. Roberts**                                   £5.95

'A stupendous achievement . . . This is the unrivalled World History for our day' – A. J. P. Taylor

# A CHOICE OF
# PELICANS AND PEREGRINES

☐ *A Question of Economics* **Peter Donaldson** £4.95

Twenty key issues – from the City and big business to trades unions – clarified and discussed by Peter Donaldson, author of *10 × Economics* and one of our greatest popularizers of economics.

☐ *Inside the Inner City* **Paul Harrison** £4.95

A report on urban poverty and conflict by the author of *Inside the Third World*. 'A major piece of evidence' – *Sunday Times*. 'A classic: it tells us what it is really like to be poor, and why' – *Time Out*

☐ *What Philosophy Is* **Anthony O'Hear** £4.95

What are human beings? How should people act? How do our thoughts and words relate to reality? Contemporary attitudes to these age-old questions are discussed in this new study, an eloquent and brilliant introduction to philosophy today.

☐ *The Arabs* **Peter Mansfield** £4.95

New Edition. 'Should be studied by anyone who wants to know about the Arab world and how the Arabs have become what they are today' – *Sunday Times*

☐ *Religion and the Rise of Capitalism*
    **R. H. Tawney** £3.95

The classic study of religious thought of social and economic issues from the later middle ages to the early eighteenth century.

☐ *The Mathematical Experience*
    **Philip J. Davis and Reuben Hersh** £7.95

Not since *Gödel, Escher, Bach* has such an entertaining book been written on the relationship of mathematics to the arts and sciences. 'It deserves to be read by everyone ... an instant classic' – *New Scientist*

# PENGUIN REFERENCE BOOKS

☐ *The Penguin Map of the World* £2.95

Clear, colourful, crammed with information and fully up-to-date, this is a useful map to stick on your wall at home, at school or in the office.

☐ *The Penguin Map of Europe* £2.95

Covers all land eastwards to the Urals, southwards to North Africa and up to Syria, Iraq and Iran * Scale = 1:5,500,000 * 4-colour artwork * Features main roads, railways, oil and gas pipelines, plus extra information including national flags, currencies and populations.

☐ *The Penguin Map of the British Isles* £2.95

Including the Orkneys, the Shetlands, the Channel Islands and much of Normandy, this excellent map is ideal for planning routes and touring holidays, or as a study aid.

☐ *The Penguin Dictionary of Quotations* £3.95

A treasure-trove of over 12,000 new gems and old favourites, from Aesop and Matthew Arnold to Xenophon and Zola.

☐ *The Penguin Dictionary of Art and Artists* £3.95

Fifth Edition. 'A vast amount of information intelligently presented, carefully detailed, abreast of current thought and scholarship and easy to read' – *The Times Literary Supplement*

☐ *The Penguin Pocket Thesaurus* £2.50

A pocket-sized version of Roget's classic, and an essential companion for all commuters, crossword addicts, students, journalists and the stuck-for-words.